Understanding Soil Change
Soil Sustainability over Millennia, Centuries, and Decades

Across the world, soils are managed with an intensity and at a
geographic scale never before attempted, yet we know remarkably
little about how and why managed soils change through time.
Understanding Soil Change explores a legacy of soil change in
southeastern North America, a region of global ecological, agricultural,
and forestry significance: from the acidic soils of primary hardwood
forests that covered the region until about 1800, through the marked
transformations affected by long-cultivated cotton, to contemporary
soils of rapidly growing and intensively managed pine forests. These
well documented records significantly enrich the science of ecology
and pedology, and provide valuable lessons for land management
throughout the world. The book calls for the establishment of a global
network of soil-ecosystem studies, similar to the invaluable Calhoun
study on which the book is based, to provide further information on
sustainable land management, vital as human demands on soil
continue to increase.

DANIEL D. RICHTER, JR. is Professor of Soils and Forest Ecology and
Co-Director of the Southern Center for Sustainable Forests at Duke
University.

DANIEL MARKEWITZ is Assistant Professor at the Warnell School of Forest
Resources at the University of Georgia.

Advance praise for this book:

"It is a grand tour of soil change at different temporal scales, done
with elegance and scientific rigor. This story will be of interest to
ecologists who have never had a soil science course, as well as to
advanced pedologists, biogeochemists, agronomists, foresters, and land
managers."
William Reiners and Pedro Sánchez

T0215475

With reference to maintaining fertility of managed soils:

"It is time ... to take stock of the concepts, data and methodologies that can be applied now or in the very short term to several troublesome questions."

Earl L. Stone (1979)

Understanding
Soil Change

Soil Sustainability over Millennia, Centuries, and Decades

DANIEL D. RICHTER, JR.
Duke University

DANIEL MARKEWITZ
University of Georgia

CAMBRIDGE
UNIVERSITY PRESS

CAMBRIDGE UNIVERSITY PRESS
Cambridge, New York, Melbourne, Madrid, Cape Town, Singapore, São Paulo

Cambridge University Press
The Edinburgh Building, Cambridge CB2 8RU, UK

Published in the United States of America by Cambridge University Press, New York

www.cambridge.org
Information on this title: www.cambridge.org/9780521771719

First published 2001
Reprinted 2002
This digitally printed version 2007

A catalogue record for this publication is available from the British Library

Library of Congress Cataloguing in Publication data

Richter, Daniel deBoucherville.
Understanding soil change: soil sustainability over millennia, centuries, and
decades / Daniel D. Richter, Jr., Daniel Markewitz.
     p.   cm.
Includes bibliographical references (p.   )
ISBN 0 521 77171 4 (hardbound)
1. Soil management – Southern States – History.   2. Land use – Southern States
– History.   3. Soil and civilization.   I. Markewitz, Daniel, 1964–
II. Title.
S591.55.U6 R53 2001
631.4′9775–dc21   00-050364

ISBN 978-0-521-77171-9 hardback
ISBN 978-0-521-03943-7 paperback

Contents

Preface

Humans are increasingly living in urban and suburban environments, away from the land and apart from the soil, yet the quality of human life and the earth's environment has never depended more on soil management than it does today. Humanity's expanding systems of food, fiber, and water production are now entirely dependent on the management practiced on several billions of hectares of soil. For these reasons, soil deserves a much greater share of human attention and affection. In the recent words of one scientist, soil is "the central processing unit of the earth's environment."

Our understanding of soil's role in the great global cycles of chemical elements lags far behind our impact on these cycles. This book argues that the management of soil at local, regional, and global scales must continue to improve, but that this improvement is limited by the notable absence of long-term soil experiments from which we can learn about how soils change through time.

This book tells the story of changes in one soil: of the genesis, degradation, and renewal of a soil on a nearly forgotten farm in rural South Carolina, USA. The farm was known for many years as the Old Ray Place, after its colorful owner, Rev. Thomas Ray (1780–1862), who lies buried in the cemetery of the local Padgett's Creek Church where he was a preacher for six decades. Many writers use the southern USA to describe particular places, characters, and events; it is a region with a long history of strife, crisis, ambiguity, and enigma. This book too explores the particulars of a southern landscape as a way to learn things that are more universal about soil, ecosystems, management, nature, and time.

This book is written for anyone interested in soil, in human dependence on soil, and in how human activities affect changes in soils and ecosystems over historical time. In universities, the material comprising the book has been read by advanced undergraduate and graduate students who reside in a variety of academic departments. At Duke University, the material has served as a source of reading and discussion for students interested in soil genesis and fertility, biogeochemistry, plant and ecosystem ecology, conservation biology, environmental history, environmental engineering and chemistry, forest and agronomic management, and sustainable development. A recommended list of readings, each matched to a chapter in the book, is found immediately after the Epilogue.

Daniel D. Richter, Jr.
Durham, North Carolina

Daniel Markewitz
Athens, Georgia

Acknowledgments

In 1957, a small group of US Forest Service researchers planted tree seedlings in two abandoned cotton fields on the Calhoun Experimental Forest, about 15 km southwest of Union, South Carolina. Little did they realize that their experimental planting would grow into a forest that would produce insights into soils and ecosystems of such interest and implication.

In 1957, land use in the southeastern United States was in transition. Cotton acreage had declined precipitously. The area of secondary pine forest was expanding. Many soils that had long been cultivated in the agricultural economy of the Old South were once again supporting forest. Reforestation of formerly plowed land grew to cover tens of millions of hectares in the region.

Since the planting of tree seedlings in 1957, the Calhoun experiment has become a significant research area for the study of pine forests in the southeastern United States. The four-decade study of soil and ecosystem change is currently the centerpiece of research on the Calhoun Experimental Forest and is the major subject of this book.

There are few plant–soil experiments in the world that possess the temporal continuity and rigorous statistical design of the Calhoun Forest Experiment. Ecologists, soil scientists, foresters, agronomists, and land managers are indebted to the foresighted scientists who gave us the opportunity to examine the long-term soil and ecosystem data presented here. Dr. Carol G. Wells, in particular, deserves special recognition for initiating the soil and ecosystem study, continuing its sampling, and assembling the soil archive from 1962 to 1982. This book is dedicated to Dr. Wells for his long-term persistence and perspective of the soil resource.

Gratefully acknowledged also are USDA Forest Service scientists, many of whom were stationed at the Calhoun Experimental Forest in the 1950s and early 1960s. In the mid-1960s, these researchers were transferred to forest science laboratories in Charleston, SC, and the Research Triangle Park and Coweeta Hydrologic Laboratory, North Carolina. These scientists have supported this long-term research in many ways, and include not only Dr. Lou Metz who organized the planting of the Calhoun experiment but also Drs. William R. Harms, Marilyn Buford, Dean DeBell, Jim Douglass, Jacques Jorgensen, and Thomas Lloyd.

Many other colleagues have contributed to the overall conduct, sampling, analyses, and interpretation of the Calhoun experiment. Most especially these include P. Heine, M. Hofmockel, B. Megonigal, and J. Raikes, for enormous commitments to outstanding chemical analyses and data presentation. Dan Binkley and three anonymous readers offered many useful suggestions in reviews of the manuscript. Some of the many others who have had an important hand in this work include H.L. Allen, R. April, W. Arthur, N. Bethala, S. Billings, W.D. Billings, B. Browne, S.W. Buol, E. Burroughs, A. Caldwell, N.L. Christensen, B.J. Cosby, C. Craft, K.H. Dai, C. Davey, E.P. Davidson, K. Davis, A.T. Davison, J. Dillon, A. Duke, J. Dunscomb, J. Edeburn, D. and J. Evans, S. Fox, C. Gnau, J. Haggard, P. Halpin, M. Hanchey, K. Hetts, C. Hoglen, Z. Holmes, E. Ingham, V. Jin, E. Jabocoby, K. Johnsen, D.W. Johnson, G. Katul, J. Kertz, K. Kipping, M.S. Knowles, K. Korfmacher, J. Krishnaswamy, I. Lepsch, S.E. Lindberg, D. Livingstone, T. Lookingbill, T. Moriya, L.A. Nelson, L.E. Nelson, B. Nettleton, R. Newton, N.-H. Oh, K. O'Neill, R. Oren, J. Powers, R. Puckett, R. Qualls, C.W. Ralston, C. Reinhart, D. Ricalde, C.J. Richardson, F. Sanchez, P.A. Sánchez, W.H. Schlesinger, D. Shoch, T. Shouse, R. Singh, P. Smith, E.L. Stone, T. Strickland, A. Stuanes, W.T. Swank, J.L. Swanson, G.L. Switzer, R. Tabachow, K. Tian, S. Trimble, S.A. Trumbore, D. Urban, J.B. Urrego, J. Vilas, M. Vinson, J. Vose, L. West, M. Williams, S. Williams, and J.C. Woodwell.

The authors also acknowledge support from resource managers of the Sumter National Forest, the US Department of Agriculture's NRI Ecosystems and Soil and Soil Biology Research Programs, the USDA Forest Service Cooperative Research Program, the National Science Foundation's Ecosystems Program, and Duke University. Duke University is home to a special community of ecosystem ecologists and to the Nicholas School of the Environment which has for more than 60 years supported research and education in soil science, forest ecology, and forest management.

Lastly, we thank the following organizations for permission to reproduce illustrations: *BioScience* (Figures 8.1 and 10.3); Carnegie

Institute (Figure 12.2); Duke University's Duke Forest (Figures 12.6 and 14.1); Duke University's Perkins Library (Figure 12.1); *Ecology* (Figures 17.1, 17.2, 17.4, and 17.5); Elsevier Science (Figures 3.2a, 9.3, 16.2 to 16.6, and 17.3); Mr. M. Hofmockel (front cover photograph of the Calhoun soil profile, and Figure 9.4); Iowa State University Press (Figure 6.6); Dr. A.T. Leiser (Figure 1.1b); *Nature*, Macmillan Magazines Ltd (Figures 15.2 and 15.3, from Richter *et al.* 1999); New York Public Library (back cover wood cut, and Figures 11.1 and 11.2); *Soil Science Society of America Journal* (Figure 6.7); Springer-Verlag (Figure 7.2); United States Library of Congress (back cover photograph; Prints & Photographs Division, FSA-OWI Collection, Reproduction No. LC-USF342-008022-A); University of Chicago Press (Figure 7.1).

The book is dedicated to Dr. Carol G. Wells, originator of the Calhoun soil-ecosystem experiment, and to four most worthy professors, Drs. Lyle E. Nelson, Charles W. Ralston, Earl L. Stone, and George L. Switzer. We also dedicate this book to our families: Susan Adam, Daniel, Christina, and Benjamin; and Jane Raikes; and parents, Nancy and Daniel, and Ahuva and Moshe.

Foreword

Historical effects often underlie otherwise puzzling observations in nature so that history is an essential element for understanding the ecological status of a place. Similarly, a historical perspective is essential for understanding soils. The great pedologist, Hans Jenny, made "time" one of a series of variables determining the state of particular soils. Ecology and soil science are parallel sciences in many ways, as well as being inextricably linked through reciprocal relationships between biota and soil condition.

Both soil scientists and ecologists seek to interpret the immediate state of ecosystems and soils (the latter considered part of the former by ecologists) through historical perspectives. For example, ecologists will attempt to understand the regrowth of a logged forest as a short-term phenomenon in the perspective of longer-term phenomena of primary succession, climate change over the span of thousands of years, migration of species, and changing status of soils. The experienced ecologist attempts to interpret vegetation dynamics in the context of soil change, but usually assumes, sometimes erroneously, that vegetation and other parts of the ecosystem's biota change more rapidly than do the underlying soils. Soil scientists are taught soil genesis in a theoretical way, but most practice their profession at time scales too short to expect change in soil formation.

Few cases exist for understanding the long-term process of soil development. The patterns of long-term change are known from the Mendocino Terraces of California, formerly glaciated terrain of Glacier Bay, Alaska, dunal terrain around Lake Michigan, and volcanic flows of Hawaii. These documented examples provide benchmarks by which we have gained insight into long-term soil development and its ecological implications. Likewise long-term agronomic research spanning decades to over one century, notably Rothamsted in the UK, Illinois' Morrow plots and shorter, but decadal ones in the tropics, has provided valuable knowledge about the chemical and physical dynamics of soils and plant growth.

Richter and Markewitz open up another such example with their sweeping treatise of the Ultisols of the southeastern USA. They give us another benchmark example for a large region having very high agricultural and forestry significance. Building on the work of earlier scientsts from decades ago, Richter and Markewitz examine soil change on the South Carolina Piedmont on multiple temporal scales: decades, centuries, and millennia. The authors carefully guide us through two interacting strands of historical narrative: pedogenesis of the old, geomorphologically stable, uplands of the southeastern Piedmont, and land-use change on the Old Ray Place, Union Co., South Carolina. The combination of these two narratives builds a fascinating story of interaction between land use and soil condition. It also leads to some important conclusions about the consequences of industrial forestry. Erosion, weathering, leaching, translocation, planters, slaves, tenant farmers, and modern foresters all play roles in this dual saga of the Old South.

This work is also of extreme importance to the tropics, even though South Carolina lies squarely in the warm temperate region. Ultisols and related soils cover vast areas of the tropics with similarly low inherent fertility, coarse-textured surface soils, and low-activity clays. For example, the main difference between most Ultisols of the study area and similarly classified ones in the tropics is the different soil temperature regime. In addition, there are strong linkages in the human influence. The history described by Richter and Markewitz is a classic saga of shifting cultivation, which was the first agricultural system in forested areas of the United States and Europe and is the prevalent system nowadays in the humid tropics, most of it on Ultisols. While few places in the tropics have gone through such a stage of agricultural intensification as has the southeastern United States, resulting in this case in millions of hectares of productive secondary forests, this book provides valuable insights on the processes involved in the transformation of slash and burn agriculture into a modern rural scene where farms are scattered in a landscape dominated by forests, which is typical of much of today's South.

Building on the long-term observations of earlier scientists, these authors show how Ultisols, a very important soil order world-wide, come into being through natural weathering, leaching, and accumulation processes. They then discuss data on the impacts of forest clearing, mixed crops and cotton farming, liming, and fertilization on these old soils. Finally, they evaluate the impacts of pine-forest growth on these old fields. The results are impressive and sometimes surprising.

This story comes to us as a result of long-term observations and systematic sampling, analyses and archiving. Unfortunately, such sustained observations are rare in the world and we must be grateful to the authors for synthesizing the data in such a palatable form. In the same vein, the authors issue a challenge to all of us. While espousing the value of synthesized, long-term studies like this, they ask why such efforts should be so rare, and whether we as a modern society concerned about long-term sustainability can commit to expansion of these scientific activities more broadly. They make a strong case for institutionalizing long-term studies at Calhoun Experimental Forest where these records were made, and for representative sites in different biomes and major soil orders elsewhere in the world.

Richter and Markewitz have combined the dedication and perspective of the early giants of soil science such as Dokuchaev, Hilgard, Kellogg, Lutz, Chandler, Jenkinson, and Jenny with modern techniques, methods, and language to produce a well woven tale of change. This tale is extraordinarily useful to ecologists and soil scientists alike as well as highly relevant to planners and managers of the New South and other parts of the world where Ultisols and similar soils underlie present and future human activities.

Not since he read the classic book *The Soil under Shifting Cultivation* (written by Peter Nye and Dennis Greenland by candlelight in Ghana in the late 1950s) has Sánchez enjoyed and learned so much about the dynamics of acid soils as from this book. It is a grand tour of soil change at different temporal scales, done with elegance and scientific rigor. This story will be of interest to ecologists who have never had a soil science course, as well as to advanced pedologists, biogeochemists, agronomists, foresters, and land managers.

William A. Reiners and Pedro A. Sánchez
June 2000

One generation passeth away, and another generation cometh:
   but the earth abideth for ever.
The sun also ariseth, and the sun goeth down, and hasteth to
   the place where it arose.
The wind goeth toward the south, and turneth about unto the
   north; it whirleth about continually, and the wind
   returneth again according to its circuits.
All the rivers run into the sea; yet the sea is not full; unto the
   place from whence the rivers come, thither they return
   again.

Ecclesiastes 1:4–7

Part I

Soil and sustainability

Managed well, soil circulates chemical elements, water, and energy for great human benefit. Managed poorly, it is impossible to imagine an optimistic future.

1

Concerns about soil in the modern world

In *Gulliver's Travels*, the ever reasonable King of Brobdingnag praised all those who toil to improve soil management (Swift 1735). The King emphasized the significance of soil management by pronouncing to a much impressed Lemuel Gulliver that

> ... whoever could make two ears of corn or two blades of grass to grow upon a spot of ground where only one grew before, would deserve better of mankind, and do more essential service to the human race than the whole race of politicians put together.

On all continents, farmers, foresters, engineers, ecologists, and gardeners labor valiantly to improve soil management (Figure 1.1). Their efforts have been amply supported by creative technicians and inventors. To harness animal power, people domesticated and bred horses, oxen, llamas, and water buffalo, and developed sophisticated yokes and collars for animal-powered hauling and plowing. To improve crop plants, maize, rice, wheat, and barley have been bred for millennia based on yields, taste, and resistance to environmental stress. To improve irrigation, Archimedes developed a simple, highly efficient screw pump. To benefit plant seedlings, Jethro Tull (1731) and many others developed seed drills. To improve soil fertility, soil amendments such as lime, organic manure, and inorganic fertilizers were promoted by John Lawes and J. Henry Gilbert (Dyke 1991), Edmund Ruffin (1852), Eugene Hilgard (1860), and Justus von Liebig (1843).

Efforts to improve soil management continue unabated in every nation. As emphasized by the King of Brobdingnag, such improvements remain among the highest priorities for humanity. How else will we feed and sustain a world which may approach 10 billion people in just a few decades?

(a, i)

(Figure 1.1)

MODERN CONCERNS ABOUT SOIL CHANGE

Soil is a biologically excited, organized mixture of organic and mineral matter; the bio-mantle of unconsolidated material that makes life possible on planet earth. Soil is created by and responds to biota, climate, geomorphic and geologic processes, and the chemistry of the aboveground atmosphere. Soil is an open thermodynamic system, highly responsive to inputs and outputs of chemical elements and energy. Although soil can be degraded, it is rarely if ever completely exhausted, due to a continuity of inputs that include solar energy, organic matter, nutrients, water, and gases. The earth's soil helps control the circulation of the biosphere's chemical elements.

(a, ii)

Figure 1.1. Vast effort is expended to maintain and improve soil for crop and forest management: (a) *opposite and above* garden cultivation and wheat harvest in Canada (Hayward and Watson 1922; photographs Edith S. Watson); (b) *overleaf* soil stabilization by vegetative wattling on steep-cut slopes in California (Gray and Leiser 1989); (c) a small farmer's sugar cane cultivation and tree planting (*Terminalia amazonia*) near San Isidro in southern Costa Rica (photograph D.D. Richter); (d) improved grass management at the Park Grass Experiment at Rothamsted in southern England (photograph D.D. Richter).

Well functioning soils are directly responsible for much of the world's highest quality freshwater, the biological diversity of terrestrial and aquatic ecosystems, and the economic wealth of human societies. Soil not only produces an increasing amount of food and fiber, but also decomposes much of the burgeoning stream of human and animal waste. Soil provides the physical support for our homes, roads, and cities (Gray and Leiser 1989). It also can be a repository for pollutants, including many of high potential toxicity.

Soil has been used for agricultural and engineering purposes for nearly 10 000 years. Over this time, humanity has developed an intimate relationship with the earth's soil that is well illustrated in our most ancient and holy texts such as the Bible, the Torah, and the Rigveda. The Rigveda speaks poetically about this relationship: "Harness the plows, fit

(b)

Figure 1.1 (*cont.*)

the yokes, now that the womb of the earth is ready to sow ..." With such a close and long-continued dependence on soil, we are assured that soil can be managed productively and sustainably in a wide variety of eco-systems.

During the 20th century, land management greatly increased soil productivity (Figure 1.2). From cereals to fuelwood, the productivity of soil burgeoned. Human consumption of protein and calories doubled between 1960 and the 1990s (Figure 1.3), from about 190 to >380 billion

(c)

(d)

Figure 1.1 (*cont.*)

grams of protein per year and from 7 to > 14 trillion calories per year. In much of the developing world, the average daily diet has gained nearly 0.5 g of protein per capita each year for the last 30 years (Figure 1.4). To achieve this production, soil inputs of N, P, and K increased 3- to 8-fold between 1960 and 1995 (Figure 1.5). Global fertilizer inputs in the 1990s totaled about 100 million metric tons (of nutrient elements) per year. In the 1990s, about 20% of the earth's arable land was irrigated, an area that nearly doubled in the final three decades of the 20th century.

Similarly, soils are being intensively managed for production of wood fiber. Industrial-wood harvests for sawlogs, veneer, pulpwood, and

(a)

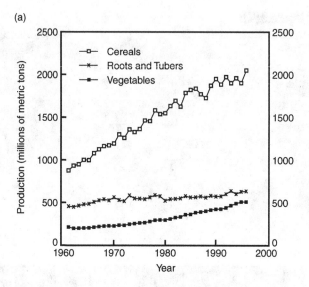

(b)

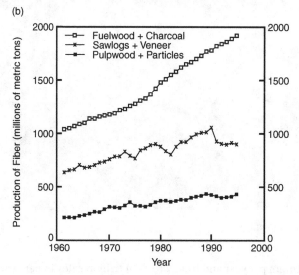

Figure 1.2. Food and fiber production are increasing rapidly and raise concerns about the soil that supports this production (UN–FAO 1998).

chips increased by 50–100% between the 1960s and 1990s (Figure 1.2b). During this same period, harvests for fuelwood nearly doubled to two billion metric tons of wood per year (Figure 1.2b).

Forest soils continue to be converted to a variety of non-forest uses. Although forestland conversion has significant uncertainties (e.g., Melillo *et al.* 1985), the Food and Agriculture Organization (FAO) reports

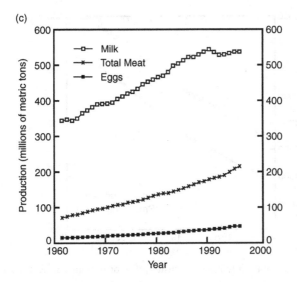

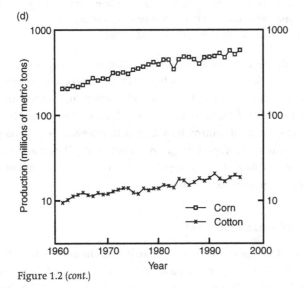

Figure 1.2 (*cont.*)

that forestland area in the 1990s is 200 million hectares less than it was in the early 1960s, mainly due to conversion to agricultural uses (Figure 1.6).

Urban impacts on soil are also intensifying as human populations are growing more rapidly in cities than in the countryside. Cities will be home to more than three billion people by 2010. To improve the quality of city life, we need to better manage soil compaction, erosion, flooding, chemical contamination, waste disposal, and runoff of polluted waters.

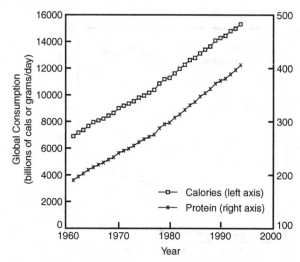

Figure 1.3. Total daily human consumption of protein and calories has increased by about 200 billion grams and 8 trillion calories between 1960 and the mid-1990s (UN–FAO 1998). Most proteins and calories consumed by humans are derived from the soil (UN–FAO 1998).

And by the end of the 1990s, mining affected on the order of a million hectares of land per year, each year creating severely disturbed landscapes in need of soil reconstruction, reclamation, and stabilization (Antonovics *et al.* 1971; Bradshaw and Chadwick 1980; Hosner and Hons 1992). Modern systems of mining have begun to integrate soil-management regimes into reclamation programs to better stabilize waste-rock dumps and tailings. Nevertheless, opportunities for soil management in mine reclamation programs remain under-developed.

HOW ARE SOILS CHANGING?

We have entered an age in which more than half of the earth's 13 billion hectares of soil are being plowed, pastured, fertilized, limed, irrigated, drained, fumigated, bulldozed, puddled, compacted, eroded, leached, mined, reconstructed, harvested, or converted to new uses. If these managed soils are not simply to degrade with use, we have much to learn about the biological fertility, chemistry, and physical stability of soil. We cannot take for granted that soil will produce ample and increasing yields of high quality food, fiber, and water over many generations' time (Greenland and Szabolcs 1994).

Gone are the days when we can readily abandon land after exploitive use and move on to "fresh soil." To preserve soil already in use, and

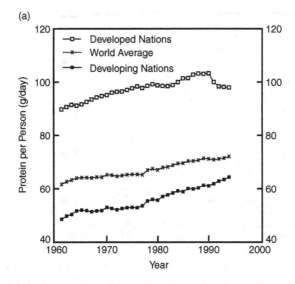

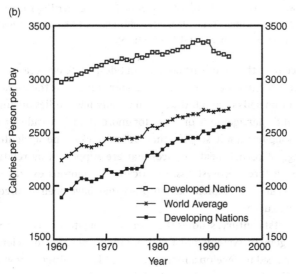

Figure 1.4. Recent global increases have been rapid in per capita, daily human consumption of (a) protein and (b) calories (UN–FAO 1998).

to improve management control over soil coming into use, we must greatly increase our technical understanding of how management alters soil over time. We have entered an age in which soil and ecosystem research is indispensable for continuing and improving land management.

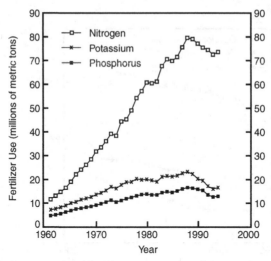

Figure 1.5. Global fertilizer consumption of elemental N, P, and K increased rapidly through the late 1980s. The increasing trend in fertilizer consumption was interrupted in the late 1980s by decreased fertilizer use in the former Soviet Union. Total elemental additions of NPK have reached 100 million metric tons per year (UN–FAO 1998).

Our central theme in this book is that despite 10 000 years of soil use we have a relatively elementary understanding about the impact of management on soils through time. Remarkably few studies document effects of land management on soils for more than a decade or two. Nearly all long-term soil studies that do exist examine agricultural systems (e.g., of corn, wheat, or rice) that are supported by relatively fertile soils. In forest ecosystems, practically no rigorous experiments exist with permanent, replicated soil plots that have been studied for more than two decades.

In the 21st century, studies that evaluate long-term sustainability of soil-management systems must become high priorities for scientific research. We need to develop a more precise understanding of how soils respond to management, and about why certain soil-management systems are sustainable. In other words, how can soil management be improved so that twin objectives of plant production and environmental quality are achieved and remain undiminished for a very long time?

Soil sustainability is not an ideal, fixed, or static goal (Lal and Stewart 1995; Nambiar 1996). Developing more sustainable soil-management systems is a practical task, a means to an end, and a process of improvement that depends directly on technical data with which we can make practical changes in the management of soils that support

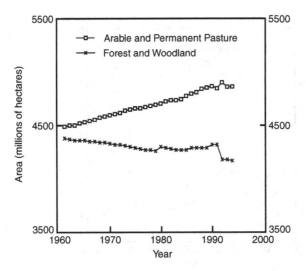

Figure 1.6. About 40% of the earth's ice-free land area (which totals about 13 billion hectares) is now cultivated land and permanent pasture. Although increases in arable land and pasture area appear to be relatively modest (about 0.22% gain per year), the expansion is enormous in absolute scale. Cultivated land plus pasture was approximately equivalent to that of forests in 1960. In 2000, cultivated land plus pastures exceeds remaining forestland by about one billion hectares. Expansion of total area in cultivation and pasture is accompanied by exploding increases in management intensity such as that used to manage irrigated cropping and permanent crops. From the 1960s to the 1990s, land under irrigation increased by about 3% per year (UN–FAO 1998).

pastures, cultivated fields, gardens, grasslands, and forests (Greenland 1994). Quantitative data from two long-term studies are illustrated in Figure 1.7 to emphasize the powerful perspectives that long-term experiments provide. In the study of wheat in the UK (Figure 1.7a), yields have varied greatly over 150 years of management but over this time have been sustained or even increased. In the study of Philippine rice (Figure 1.7b), production has declined in these fields and little adequate explanation yet provided (Cassman *et al.* 1995).

PURPOSE OF THE BOOK

This book is interested in how all soils change through time: how soils are formed by pedogenesis, affected by a land-use history, and are currently changing in modern ecosystems that have an increasing human influence. Although our broad subject is soil change through time,

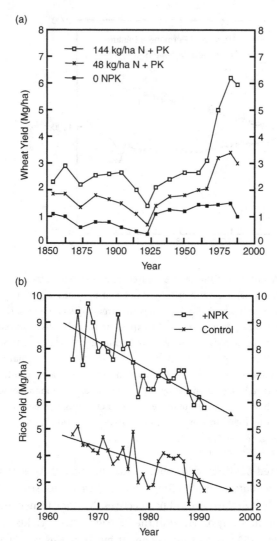

Figure 1.7. Yields of (a) winter wheat grown on the Broadbalk Field, Rothamsted, UK from 1852 to 1990 (Johnston 1994), and (b) irrigated rice at the International Rice Research Institute (IRRI) in the Philippines from 1964 to 1990 (Cassman and Pingali 1994). These experiments demonstrate the possibility and challenge of maintaining high crop yields. Data from Broadbalk illustrate effects of the introduction of weed control (in the 1920s with fallows and in the 1960s with herbicides), fungicides (1980s), and eight new wheat cultivars. Rice data from IRRI illustrate high but declining yields for dry season crops over two decades. Causes for these dramatic declines remain undetermined according to Cassman and Pingali (1994), but include the possibility of pathogens, nitrogen deficiencies, organic matter interactions, and water quality.

Table 1.1. *Because soils are open systems, they proceed through developmental stages of weathering as energy, water, and chemical elements are processed. Three general weathering stages were used by Jackson and Sherman (1953) to illustrate this genesis of soils; we illustrate the implications for soil orders and common soil minerals*

	Jackson–Sherman (1953) soil weathering stage		
Attribute	Early	Intermediate	Advanced
Soil Taxonomy orders (Soil Survey Staff 1998)	Entisol, Andisol	Inceptisol, Mollisol, Alfisol	Ultisol, Oxisol
Common soil minerals	Gypsum, calcite, olivine, biotite, feldspar	Feldspar, muscovite, vermiculite, smectite	Kaolinite, gibbsite, hydrous Fe oxides

our special interests are temporal changes that occur in acidic, advanced weathering-stage soils (Table 1.1), soils such as those that dominate not only southeastern North America but also an enormous part of the tropics and warm temperate zone as well.

To use soil without damaging it is always a difficult challenge, but management is especially problematic with advanced weathering-stage soils, since they may lack components and processes that can buffer impacts of management. Understanding change in advanced weathering-stage soils is vitally important to maintaining the world's land management already in place, and also because extensive areas of these soils support tropical forests that are currently being cleared and converted to agricultural uses.

We explore the story of soil change on one farm in the South Carolina Piedmont: how this farm's soil formed by pedogenic forces, was transformed by a century and a half of agricultural cultivation, and is being changed again by recent decades of forest growth and development.

The specific objectives of this book are to evaluate soil's ability to sustain its fertility and specifically its plant-nutrient supply over time scales of millennia, centuries, and decades. The objectives thus correspond to soil change over these three time scales, and are addressed in the book's three main parts which evaluate:

1. *Soil genesis or the natural processes of soil formation that over millennia produce advanced weathering-stage soils (Table 1.1).* Across much of the earth's surface, natural soil formation and ecosystem development may accumulate organic carbon and nitrogen, but may lead also to diminished soil fertility, soil acidification, and the depletion of many nutrients.

2. *Land-use history that over time scales of centuries has both enriched and degraded native soil fertility.* We can learn a great deal about soil change by examining the legacies of agriculture in the southeastern USA, a region where soils and ecosystems that have been transformed by previous agricultural use.

3. *Current ecosystem development that over time scales of decades is altering soil fertility, especially in intensive systems of land management.* Across much of the southeastern USA, formerly agricultural land has now been reforested, a process that is often thought to benefit soils and soil fertility. Relatively few data precisely document effects of forest growth on soil biogeochemistry (Nye and Greenland 1960; Stone 1975; Sánchez *et al.* 1985; Fisher 1990; Johnson *et al.* 1991; Evans 1992, 1994; Binkley and Giardina 1997; Alriksson 1998). Much can be learned about the dynamics of soil change at the Calhoun Experimental Forest, one of the world's longest-running studies of soil change with rigorous experimental design and soil archive.

These are exciting times for those with passionate interests in the earth's soil, and the ability of soil to support human societies and natural ecosystems. A high priority for the 21st century is similar to that stated by the King of Brobdingnag and Jonathan Swift in the early 18th century: to improve soil management on all continents in our attempt to keep pace with rising human demands for a better life.

2

Managing soils for productivity and environmental quality

"Moving up the yield curve" has dominated much of the modern approach to soil management (Figure 2.1). To maximize crop production, soil management has often been narrowed to a dose–response relationship between technical inputs and harvest outputs.

Dose–response relationships guide soil management for corn, rice, wheat, beans, pulpwood, cotton, bananas, pineapples, coffee, potatoes, and even road-side vegetation planted to stabilize slopes. Typical fertilizer-dose recommendations include:

- yearly additions of nitrogen at 75 to 150 kg ha^{-1} to high-yielding rice in Indonesia, China, India, and the Philippines;
- yearly additions of nitrogen at 200 to 300 kg ha^{-1} to intensive, shade-free coffee in Costa Rica;
- yearly additions of potassium at 400 to 700 kg ha^{-1} to bananas in the humid tropics of Central and South America;
- periodic additions of nitrogen at 200 and phosphorus at 50 kg ha^{-1} to pine forests in southeastern North America; and
- yearly additions of nitrogen at 50 to 150 kg ha^{-1} to residential lawns and park grass throughout the world.

The objective of these management regimes is straightforward: to add sufficient nutrient input to overcome soil limitations and maximize the profitable growth of plants. Land managers on all continents, whether farmers, foresters, or engineers, are faithfully following dose–response recommendations in an effort to help crop plants achieve high yields and profits.

There is little doubt that the dose–response approach to soil management can be credited with much of the global increase in crop

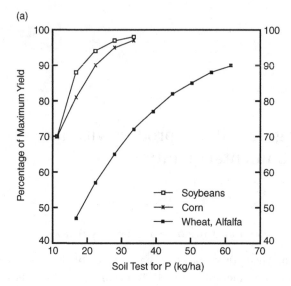

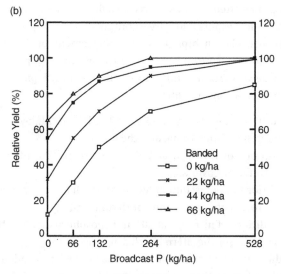

Figure 2.1. The basis for the dose–response approach to soil management can be illustrated by fertilizer–crop yield relationships. (a) The yield response of soybeans, corn, wheat, and alfalfa to concentrations of phosphorus in soil. Maximum achievable yield is 100%. Recommendations for technical inputs have often been based on a target of 90% of maximum potential yield. (b) Crop-yield response to method of phosphorus application on a phosphorus-deficient soil. Fertilization experiments compared phosphorus inputs across the entire field (broadcast) with that applied in bands directly in plant rows. (c) Results of a simple fertilizer dose–response experiment for four crops (Sánchez 1976).

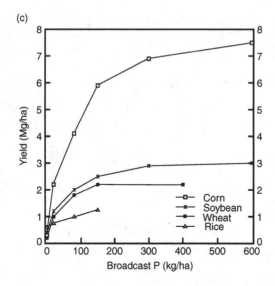

(c)

harvests during the 20th century. These patterns in plant production are spectacular (Figures 1.2 to 1.5), and are in many ways amongst the most impressive achievements of modern civilizations.

From a broad perspective, however, the dose–response approach has also brought serious environmental problems. Some of the most adverse are off-site effects on water resources. Recent water quality studies clearly demonstrate how extensively modern fertilization affects water resources (Figure 2.2). Large fractions of the nutrients that are transported in the world's rivers, lakes, and estuaries are derived from leakage of nutrients from fertilized ecosystems. Modern agro-ecosystems can be characterized by their relatively inefficient use of applied fertilizers.

Broad concern over water quality and soil productivity illustrate why the purview of soil management must be broadened far beyond dose–response relationships of technical inputs and harvest outputs. "Moving up the yield curve" will always be important to managed ecosystems, but soil management must more actively minimize adverse on-site and off-site effects as well. The goal for soil management in the 21st century is a system that co-values crop production and high environmental quality.

OPPORTUNITIES TO IMPROVE SOIL MANAGEMENT

We have many practical opportunities to improve soil management so that yields will come with fewer adverse on-site and off-site effects. We

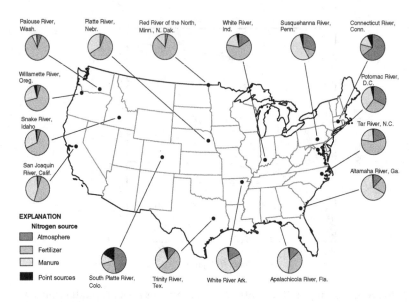

Figure 2.2. Sources of nitrogen in natural waters of selected river basins of the USA. In this study, Puckett (1994) estimated that nitrogen in most rivers studied was derived from fertilizer and manure sources.

can establish greater management control over the cycling of fertilizer nutrients, organic matter decomposition and quality, the activity of biological N_2 fixation and rhizosphere microflora, and the depth of plant rooting. We can develop more sophisticated soil and plant analyses, better account for nutrient deposition from the atmosphere, expand use of crop rotations with multiple purpose crops, improve control over the solubility of chemical fertilizers, improve placement and timing of fertilizer and other chemical applications, and even use variable-rate applications of fertilizers within individual management units.

Improving soil-nutrient management is not a new idea. In the 1840s, soil-management regimes were evaluated to determine whether manure could be applied to fields throughout the year or would best be saved for the early growing season when plant uptake of nitrogen was highest and drainage loss potentially reduced (Way 1850). In the 1880s, nutrient-retention efficiency was tested in studies of the effect of fertilizer timing on nitrate leaching (Table 2.1).

Nutrient-retention efficiency of agro-ecosystems has also been estimated in an on-going 100-year-old study of crop rotations in Alabama, USA (Table 2.2). Soil-management treatments in this study compare the fate of nitrogen from fertilizers and from leguminous plants.

Table 2.1. *Nutrient-retention efficiency has been a topic of research for well over a century (Johnston 1994). In 1878–1883, nitrate (mg L^{-1} as nitrogen) was estimated in water draining from nitrogen-fertilized wheat plots at Broadbalk, Rothamsted. Fertilizers (NH$_4$SO$_4$) were mainly spring applications to benefit growth of plants in summer, except one fertilizer treatment that was in autumn. Based on the concentration data, researchers concluded that efficiency of fertilization was much higher in spring due to high plant uptake during the growing season*

		Nitrate-nitrogen concentration in drainage water				
Fertilizer (kg N ha^{-1})	Season of application	March to May	June to Harvest	Harvest to November	November to March	Yearly average
				(mg L^{-1})		
0	—	1.7	0.2	5.6	4.5	3.9
48	Spring	8.1	0.7	7.3	4.8	5.0
96	Spring	16.3	1.4	8.3	5.2	6.4
96	Fall	5.7	2.9	7.4	26.4	19.4
144	Spring	21.5	4.0	14.7	7.3	9.3

Results indicate that crop harvests contain 20 to 72% of nitrogen inputs (from nitrogen fertilizer, N$_2$ fixation, and atmospheric nitrogen deposition), indicating a large potential for management to affect nutrient retention.

Some agronomists and ecologists view modern improvements in soil management with impatience, as it evolves from a dose–response approach to one that more explicitly values both productivity and environmental quality (Rodale 1945; Jackson 1980; Berry 1995). Sánchez (1994) calls for a new "soil-management paradigm," one that is based on managing the full nutrient cycle rather than only fertilizer inputs, and on an approach that relies more on biological processes and less on technological practices to overcome soil limitations.

The Sánchez (1994) approach is less a new paradigm, at least as defined by Thomas Kuhn (1970), than a well articulated perspective of soil management that is jointly aimed at both productivity and environmental quality. It is not too much of a stretch to point to a long line of persons who have worked in parallel with the Sánchez paradigm. These include:

Table 2.2. *Six-year nitrogen budgets for five cotton-cropping systems in the Old Rotation study in Alabama, USA (Mitchell et al. 1996). About 20 to 70% of nitrogen input (from fertilizer, N_2 fixation, plus atmospheric nitrogen deposition) is recovered in crop harvests from these cropping systems*

| | N inputs from | | N in crop | Harvest |
| | Legume | Fertilizer | harvests | efficiency of |
Cropping system	$-$(kg ha^{-1} 6 years^{-1}) $-$			N input[a] (%)
Continuous cotton with no N inputs	0	0	72	—
Continuous cotton with N fertilizer	0	720	240	32
Continuous cotton with winter legume	696	0	228	31
Cotton–corn rotation with winter legume	692	0	238	33
Cotton–corn rotation with winter legume and N fertilization	692	716	280	19
Cotton–corn–soybeans with winter legume and winter rye	640	120	550	70

[a]Additional nitrogen input is derived from atmospheric deposition which totals about 30 kg N ha^{-1} over the six-year rotation period. Harvest efficiency of nitrogen input is expressed as a fraction of total input (legume, fertilizer, and atmospheric deposition). Seed-cotton yields averaged 930, 1860, 2230, 2290, 2560, and 2240 kg ha^{-1} year^{-1} for the six cropping systems, respectively.

- Xenophon in ancient Greece and Junius Columella in Rome, who were devoted to improving the quality of soil management, especially by promoting use of legumes and nutrient amendments.
- E.W. Hilgard (1860), a "father" of soil science (Jenny 1961b), who traveled on horseback in the southern USA in the 1850s to sample and analyze the region's ecology, geologic substrata, and soils. Hilgard (1860) was interested in matching soil capability with land use.
- Charles Darwin (1897), a forgotten "father" of soil science (Paton *et al.* 1995), who was fascinated by the rates at which earthworms regenerate soil.

- George P. Marsh (1864), Mahatma Gandhi (1940), and Aldo Leopold (1949), who were as concerned about sustaining human uses of soil as they were about ethical dimensions of land management.

By the late 20th century, many people had helped develop concepts and technical information needed for a broadly based soil management that is not dissimilar to that advocated by Sánchez (1994). These persons have combined interests in soil as a biological system with a practical perspective that billions of people depend on soil to grow food and fiber and to provide an income. Some who have been as interested in the details of ecosystem process and function as in crop yields include microbiologists (Winogradsky 1938), chemists (Jackson 1964), soil-fertility specialists (Black 1968; Blackmer *et al.* 1991), ecologists and foresters (Lutz and Chandler 1946; Wilde 1958; Ovington 1962; Stone 1975), and analysts interested in the circulation of nutrients in natural and managed systems (Rennie 1955; Miller *et al.* 1978; Bornemisza 1982; Jenkinson 1991).

Most assuredly, the King of Brobdingnag was correct in his assessment of his subjects who were engaged in important service to the human race.

3

Biogeochemical sciences in support of soil management

ORIGINS OF BIOGEOCHEMISTRY

In the late 20th century, the science of biogeochemistry emerged from its roots in geology, ecology, soil science, limnology, agronomy, and forestry, to become a major interdisciplinary science (Hutchinson 1957; Likens *et al.* 1977; Jenny 1980; Buol *et al.* 1989; Gorham 1991; Schlesinger 1997; Barnes *et al.* 1998; Burke *et al.* 1998). The scope of the new science includes all chemical reactions of the earth's surface, all living organisms and their abiotic environments. The science has much to contribute to modern soil management.

Fundamental ideas and concepts of biogeochemistry were first formulated by 19th century agronomists, geologists, and foresters. The clearest examples come from Europe. The German forester, E.W.H. Ebermayer (1876), first quantified the circulation of nutrients in forests. Specifically, Ebermayer evaluated whether litter raking and tree harvesting would significantly interrupt the nutrient cycle and adversely affect forest productivity. Among geologists, J.J. Ebelmen (1845) and K.G. Bischof (1847) examined the role of CO_2 released by biota that promoted soil-mineral weathering and the release of soluble nutrients (Berner and Maasch 1996). Ebelmen and Bischof combined laboratory and field observations to describe sources of CO_2 and sinks of O_2 with uncanny accuracy. At Rothamsted Experimental Station in Harpenden, England, agro-ecological studies began in full force in the 1840s. By the mid-1860s, agricultural fields at Rothamsted were being used as experimental ecosystems for measurements of long-term productivity, crop-nutrient uptake, and nutrient-retention efficiency. Nutrient budgets of whole fields, soils, and crops were estimated for various management treatments. From today's perspective, these and other 19th century studies developed and applied an early concept of the ecosystem.

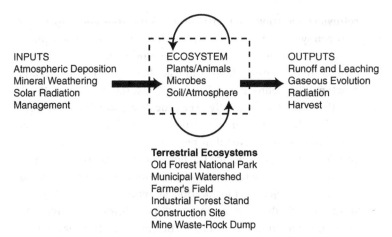

INPUTS
Atmospheric Deposition
Mineral Weathering
Solar Radiation
Management

ECOSYSTEM
Plants/Animals
Microbes
Soil/Atmosphere

OUTPUTS
Runoff and Leaching
Gaseous Evolution
Radiation
Harvest

Terrestrial Ecosystems
Old Forest National Park
Municipal Watershed
Farmer's Field
Industrial Forest Stand
Construction Site
Mine Waste-Rock Dump

Figure 3.1. Ecosystem diagram including input and output fluxes. Examples of ecosystems are farmers' fields, forest stands, grazed grasslands, national parks and wilderness areas, wetland preserves, municipal watersheds, road-construction sites, and the tailings and waste-rock dumps of open-pit mines.

BIOGEOCHEMISTRY OF SOILS: A PRIMER

In recent decades, enormous amounts of biogeochemical data have been collected that describe the functioning of a wide variety of soils and ecosystems. The ecosystem is well recognized to be the fundamental unit of ecology (Evans 1956), a dynamic, non-equilibrium system composed of biota and an abiotic environment (Figure 3.1). The rates of many ecosystem processes are now well quantified: photosynthesis, microbial decomposition, nutrient mineralization, and plant-root uptake of nutrients. Details of the hydrologic cycle have been estimated in many of the world's terrestrial ecosystems with seemingly innumerable studies of precipitation, canopy interception, infiltration, runoff, and evapotranspiration. Soil biogeochemical reactions have been studied in ecosystems as different as cultivated row crops, pastures, primary and secondary forests, savannas, wetlands, agroforestry systems, air-pollution affected forests and fields, shifting cultivation systems, and forests and grasslands managed for watershed and reclamation objectives.

This work indicates that soil development and the state of chemical elements within soils are controlled by four major biogeochemical processes: inputs, translocations, transformations, and outputs (Simonson 1959).

• *Input processes* include atmospheric deposition, biological N_2 fixation, nutrient release during mineral weathering, alluvial and colluvial deposition, and fertilization. Since soils and ecosystems are open systems, inputs of nutrient elements are critical to maintaining soil-nutrient supply over time. However, inputs are often difficult to measure with precision (Lindberg et al. 1986; April and Newton 1992; Johnson and Lindberg 1992; Galloway et al. 1994), and they can change greatly over time. For example, input from periodic volcanic explosions or from excreta of migrating flocks of birds can be massive even if infrequent.

- *Translocation processes* include the movement of chemical elements within the soil and across the landscape. Solutes, water, solids, and gases are all subject to translocation. Materials translocated downward include drainage water and its solutes, clay micelles, hydrous oxides of iron and aluminum, and gases. Materials can be translocated upward within soils in response to capillary movement of water, animal mixing, treefall disturbance, plant-root uptake, and gaseous diffusion. Recycling of chemical elements between plants and soil is also a critical translocation flux.

- *Transformation processes* include organic matter decomposition and humification, redox reactions, desilication, hydration, and mineral weathering and re-synthesis. The weathering reactions of primary minerals that were inherited from geologic material often involve the synthesis of secondary crystalline minerals and amorphous compounds in incongruent reactions. Chemical elements are also transformed in soils by adsorption of soluble ions on exchange sites (by cation and anion exchange), by incorporation of chemical elements in soil organic matter via microbial immobilization and metal complexation, or by incorporation of nutrient elements within secondary minerals (e.g., potassium "fixation" in hydroxy-interlayered vermiculite, or HIV).

- *Removal processes* include hydrologic leaching and runoff; wind and water erosion; fire-caused oxidation, volatilization, and particulate losses; gaseous emissions; and crop harvests. Since soils are open systems, soil fertility depends on a long-term balance of removals with inputs. Although soils rarely become completely exhausted of materials and energy, removals that outpace inputs eventually lead to soil depletions. Like many input processes, removal processes are difficult to measure with precision, and thus rates of soil depletion often have high uncertainty.

Over time, these four sets of processes operate simultaneously and common patterns of soil change can be observed (Simonson 1959). Such patterns provide a framework for understanding soil biogeochemical change. Figure 3.2 summarizes some possible patterns of change over long pedogenic time scales in phosphorus, carbon, nitrogen, calcium, and potassium, changes that suggest major processes controlling soil change.

The pattern of soil phosphorus illustrated in Figure 3.2a is derived from Walker and Syers (1976) who summarized a variety of observations of soils, mainly in New Zealand. After relatively rapid weathering release and decomposition loss of calcium phosphate in primary minerals such as apatite, part of the weathered phosphorus is accumulated in organic matter and in inorganic phosphorus compounds of various solubility. As soil weathering is advanced, mainly organic phosphorus and occluded inorganic phosphorus resist further loss (Figure 3.2a).

The pattern of soil change in carbon and nitrogen illustrated in Figure 3.2b is by no means universal but is one from a soil development sequence in Hawaii (Torn *et al.* 1997; Crews *et al.* 1995). There the trajectory of soil-mineral weathering strongly controls sequestration of soil organic matter. Initial formation of high-activity, organophilic minerals (allophane, imogolite, and ferrihydrite) stabilizes considerable amounts of organic matter that was added by vegetation. Such clay minerals adsorb organic matter strongly and protect it from microbial decomposition. As high-activity minerals eventually destabilize and weather to more crystalline minerals (kaolinite, gibbsite, goethite, and hematite), soil organic matter becomes susceptible to microbial decomposition. In the Hawaiian systems, organic matter declined by nearly 50% in the advanced weathering-stage soils (Crews *et al.* 1995).

Patterns of soil change in potassium and calcium illustrated in Figure 3.2 are derived from soil observations throughout the world. Weathering releases both elements to solution, leaving them potentially susceptible to hydrologic leaching. For calcium, few secondary minerals can buffer losses and total calcium eventually dwindles and may even approach exchangeable contents in the most advanced weathering stages of soil. In contrast to calcium, potassium can be effectively retained within soils by incorporation into a variety of secondary minerals, even in advanced weathering-stage soils.

A CRITIQUE OF BIOGEOCHEMISTRY

The current state of the biogeochemical sciences allows us to ask, but not to answer, significant questions about how managed soils and associated ecosystems will change in the foreseeable future. Mainly this

(a)

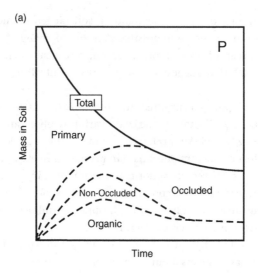

(b)

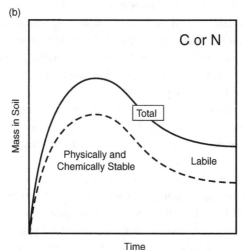

Figure 3.2. Change in forms and amounts of five soil elements with pedo-genic time (after Walker and Syers 1976; Crews *et al.* 1995; Torn *et al.* 1997). Such biogeochemical patterns can provide an understanding of trajectories and processes that affect soil change.

serious deficiency is attributed to the notable absence of soil-ecosystem studies with which we can directly observe soil changes over time scales of decades.

Although long-term ecological studies are widely valued (Likens 1989; Risser 1991), the study of long-term soil change has been notably lacking. Decades-long studies of vegetation dynamics, stream-water

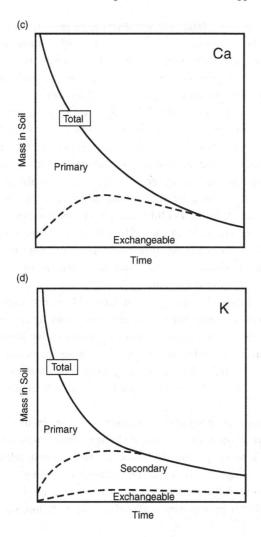

chemistry (Likens *et al.* 1977), and hydrology (Swank and Crossley 1988) have made fundamental contributions to science, but remarkably few soil-ecosystem experiments provide comparable observations of soil-chemical change over these time scales. This book describes one such soil-ecosystem experiment, located at the Calhoun Experimental Forest, in which changes in the biogeochemistry of soils and ecosystems have been directly observed over four decades.

Ironically, enormous amounts of biogeochemical data have been collected that describe soil and ecosystem processes as they operate over one to a few years (e.g., Schlesinger 1997), but nearly all of these studies are limited in their ability to project changes in soils and ecosystems

over time scales of decades. Relatively small errors in short-term estimates when extrapolated over decades time can readily bias our view of
the trajectories of soil and ecosystem change. Model simulations, no
matter how sophisticated, provide no substitute for direct observations
of change in soils and ecosystems over time scales of decades.

Currently, there is great opportunity for long-term soil-ecosystem
experiments such as that at the Calhoun Experimental Forest to make
important contributions to our understanding of biogeochemistry, ecosystem ecology, forestry, and agronomy. By directly observing the
change of soil and whole ecosystems over time, the integrated effects of
individual processes (of inputs, translocations, transformations, and
removals) can be observed, and difficult-to-measure processes evaluated.
Such soil-ecosystem experiments can provide a special perspective of
carbon turnover, nitrogen accretion, mineral weathering, aluminum
mobilization, and sulfate sorption. Both simulation models and management can benefit greatly from such data.

The science of biogeochemistry has an incredibly rich tradition
spanning several hundred years and a number of scientific disciplines
(Burke *et al.* 1998). Based on research questions currently under investigation, biogeochemistry is in a dynamic stage of its development. This is
a science that has ambitions for quantifying biogeochemical cycles
across regional and global scales, and across geologic time (Schlesinger
1997).

Some of the most exciting and significant questions in modern
ecology are questions about biogeochemical change through time. In a
world that may soon be supporting 10 billion people, understanding,
predicting, and managing soil and ecosystem change are high-priority
challenges, to say the least. The next chapter examines methods that are
used to estimate soil change over time scales of decades to millennia.

4

The science of estimating soil change

Despite a general notion that soils are relatively static, well buffered components of terrestrial ecosystems, they are increasingly recognized to be dynamic systems capable of changing in many properties quite substantially over time scales of decades.

The temporal dynamics of the soil system indicate that special approaches need to be used to observe change through time. Two approaches commonly used include (1) space-for-time substitutions (Pickett 1989), sometimes referred to as chronosequence studies, and (2) long-term soil-ecosystem studies (Leigh and Johnston 1994), based on direct observations of soil change in permanent field plots.

This book combines both approaches to evaluate how and why soils change over time scales ranging from decades to millennia. Before we examine soil change across these time scales, however, some strengths and weaknesses of these two approaches need careful consideration.

SPACE-FOR-TIME SUBSTITUTIONS

Changes in soil and ecosystems that happen over decades occur at a pace that is relatively slow in relation to human events. Initiating experiments to examine such phenomena is often not practical. If an experiment will take half a human lifetime, what happens if the experiment should fail? Moreover, scientific studies often gain value by being conducted expeditiously. Pressing issues must be addressed; university degrees must be earned. It is no wonder that for many years ecologists have almost exclusively used the time-efficient space-for-time substitution to evaluate soil and ecosystem change through time.

Long before ecologists used the space-for-time approach, Tocqueville (1838) eloquently described the procedure in his studies of the westward-expanding American society in the early 19th century. Tocqueville (1838) traveled across the western frontier and "counted on finding the history of the whole of humanity framed within a few degrees of longitude." Instead of directly observing the same town, village, or farm change through time, Tocqueville selected geographically arranged communities to represent temporal stages of system development.

Not only is the space-for-time approach expeditious, but it also provides the only practical means to understand soil and ecosystem change over centuries and millennia. Two particularly informative studies that explore soil change over Pleistocene times (on the order of two million years) are of soils of the Mendocino staircase on the northern coast of California (Jenny 1980) and of soils on the Hawaiian islands (Chadwick et al. 1999). We examine these studies briefly in Chapter 7. Other space-for-time studies that have expanded our understanding of soil change over long sweeps of time include those of soil organic matter accumulation in mudflows on the slopes of Mt. Shasta (Dickson and Crocker 1953, Sollins et al. 1983), organic matter accumulation in old-field forests (Schiffman and Johnson 1991), changes in nitrification and denitrification in New England plant successions (Melillo et al. 1983; Thorne and Hamburg 1985), and effects of primary succession following deglaciation and floodplain development in Alaska (Walker et al. 1986; Chapin et al. 1994). Other sites in which space-for-time substitution can be used to investigate soil genesis and change include volcanic ashfall deposits, glacial moraines, lava flows, river alluvium (Harden 1988), beach deposits along lake shores (Franzmeier and Whiteside 1963), soils under trees of great age (Zinke and Crocker 1962), and in organic soils periodically burned by wildfires (O'Neill 2000; Richter et al. 2000a).

When the space-for-time substitution is used to study soil change, it is often called a chronosequence study (Dickson and Crocker 1953; Jenny 1980; Sollins et al. 1983; Harden 1988). According to Jenny (1980), a soil chronosequence is a series of soils which develop on similar "landscape positions that have comparable state factors, save age t". In a soil chronosequence, the assumed-to-be constant soil-forming factors include biological organisms, climate, geologic substrata, and topography. These are often large assumptions, to say the least.

Soil scientists and ecologists including Buol et al. (1989) and Pickett (1991) have been skeptical about the wide applicability of chronosequence approaches to studies of ecological change. After all, chronosequence results may be misleading or simply false if space and time were not freely substitutable (Pickett 1991). While crediting Jenny's (1980) work with chronosequences, Buol *et al.* (1989) also wrote, "It is exceedingly difficult to impossible to ... ensure comparability of the factors of [soil] formation." The chronosequence approach is an indirect method for studying ecological change, and by definition it entirely confounds space and time. This entanglement is both the method's strength and its weakness. Only by accepting the basic critical assumption of space-for-time substitution, that all "state factors save age t" are constant, can the space–time entanglement be readily unraveled.

Although we certainly do not reject the method of space-for-time substitution, we do prefer to conceive of this approach as a soil comparative study, and require the investigator to explicitly consider all interactive pedogenic factors prior to making conclusions about the main effects of time.

Soil comparative studies can be invaluable to understanding soil change, but only if sites are very carefully selected and assumptions explicitly considered when making interpretations. Such studies are instrumental in answering otherwise inaccessible questions about trajectories of change in soils or ecosystems as they occur over centuries and millennia.

LONG-TERM SOIL-ECOSYSTEM EXPERIMENTS

A more direct approach to understanding soil and ecosystem change is that of the field experiment in which changes in soils and ecosystems are directly observed (Tinker 1994). The ultimate duration of such field experiments is limited for practical reasons, since time must elapse before results are useful. The oldest long-term soil study dates from the mid-19th century, although most long-term field experiments are several decades in age.

Long-term experiments are not easy to initiate or to sustain. Experiments must be well organized; field plots and individual samples must be well replicated; soil and ecosystem samplings must obtain representative collections. Sample archive and data management systems must be of exceptional quality. It is not surprising that we have inherited so few well replicated, decades-long soil and ecosystem studies.

In the mid-19th century, European and North American agronomists initiated long-term field research to evaluate changes in soil fertility and plant productivity that would occur with different management systems. The best known of these studies originated in the 1840s at Rothamsted Experimental Station in southern England, when John Lawes and J. Henry Gilbert initiated research that continues to this day.

Table 4.1. *Summary of the world's 12 soil orders and diagnostic horizons. Degree of weathering and soil development generally increases downward through the table. Nearly all long-term soil-ecosystem studies are conducted on Mollisols, Alfisols, and Vertisols, arable soils with high native fertility. Ultisols and Oxisols, some of the earth's most weathered and acidic soils, are included in notably few long-term soil-ecosystem studies*

Soil order	Typical horizons	Brief description
Initial weathering-stage soils		
Entisol	O, A, C	Youthful soil with little profile development
Andisol	O, A, (B)[a], C	Non-crystalline allophane clay with prominent organic-rich A horizon
Gelisol	O, A, (B), C	Permafrost within 1 m of soil surface
Histosol	O, C	Organic accumulations of various depths and chemistry, due to anaerobic conditions
Intermediate weathering-stage soils		
Inceptisol	O, A, B, C	Inception of soil horizonation and weathering (especially in the B horizon)
Aridisol	O, A, B, C	Arid to semi-arid ecosystems of shrub and short grasses
Vertisol	O, A, (B), C	High-activity, expansible clays predominate
Mollisol	O, A, (B), C	Semi-arid or moist grassland ecosystems with deep organic-rich A horizon
Alfisol	O, A, B, C	Clayey (argillic or kandic) B horizon with low acidity
Advanced weathering-stage soils		
Spodosol	O, A, (E), B, C	Humic B horizon with high acidity and coarse texture
Ultisol	O, A, (E), B, C	Clayey (argillic or kandic) B horizon with high acidity
Oxisol	O, A, B, C	Oxic B horizon with low-activity clay

[a]Horizons in parentheses are possible but not necessary to the soil order.

Effects of crop management on soil properties and processes have been tested on some Rothamsted plots for up to 150 years (Figure 4.1), including crops such as winter wheat, barley, root-crops, and hay. In several of these studies, soil and plant samples have been collected and archived for over a century. With little question, the Rothamsted studies provide the world's longest scientific records of soil and ecosystem sustainability.

Many significant observations have been made in the Rothamsted studies. Figure 1.7 illustrates long-term winter wheat yields at Broadbalk Field. As a result of these data and others like them, Lawes and Gilbert were knighted for their work and they are credited with demonstrating that, at least under Rothamsted conditions with Alfisol soils (Table 4.1),

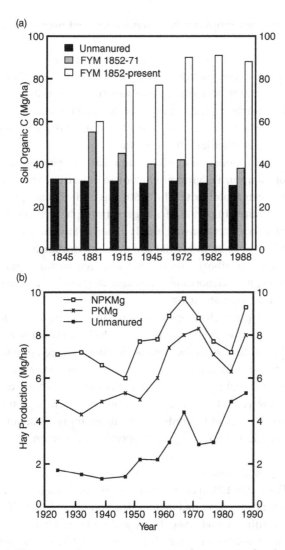

Figure 4.1. Long-term ecosystem data accumulated at Rothamsted Experimental Station in southern England. Data from Rothamsted experiments represent some of our best observations of sustainability of agricultural soil and ecosystems. (a) Changing contents of soil organic matter in response to long-term manuring treatments at Rothamsted (Jenkinson 1991). The treatment FYM is farmyard manure. The plant crop is wheat on the Broadbalk Field, and soil depth sampled is approximately the upper 20 cm of mineral soil. (b) Long-term hay yields that illustrate upturns in production even in unfertilized plots in recent decades, in part attributed to a plant-growth response to increased atmospheric inputs of nitrogen derived from pollutant sources.

yields from continuous crops can be sustained for well over a century (Figure 1.7).

In related studies at Rothamsted, soil-carbon accretion and turnover have been the focus of study (Jenkinson and Rayner 1977; Jenkinson 1991). During a century of applying farmyard manure, soil organic matter (SOM) has nearly tripled. On other plots, mineral soil that had been manured yearly between 1851 and 1871 still appears to have elevated SOM over a century later (Figure 4.1a).

Long-term hay production at Rothamsted has also been informative (Figure 4.1b). An upturn of grass production in unfertilized plots and in the fertilizer-minus-N plots after the 1950s was entirely unexpected. The pattern is attributed in part to increases in atmospheric deposition of pollutant nitrogen as Rothamsted is only about 20 km north of London and is surrounded by an industrial and urban environment that in recent decades has contributed substantial nitrogen to the regional atmosphere. Atmospheric deposition of nitrogen may have totaled up to 50 kg ha^{-1} year^{-1} following about 1960.

Such soil and plant changes can be documented in only a few long-term studies anywhere in the world, and such changes are not at all able to be studied with a space-for-time or chronosequence approach. Without long-term records, changes in soils and plant dynamics remain in what Magnuson (1990) calls the "invisible present." In other words, although we are aware that soils and ecosystems change through time, without long-term studies and records we have little ability to explain or to judge the biological significance of temporal variations in soils or ecosystems.

Fortunately, Rothamsted is not the only site with long-term soil-ecosystem studies (Table 4.2). In an Australian study initiated in 1925, 12 systems of crop rotations have had major effects on the productivity of wheat and soil fertility (Russell 1960; Grace and Oades 1994). Rotations with legumes have helped maintain high wheat yields, soil organic matter, and nutrient supply. The 38-year gains of nitrogen under leguminous pastures amount to accretions of 8.3 kg ha^{-1} year^{-1}. Fallow treatments generally degraded soil fertility and productivity.

In Missouri and Oklahoma, USA, two studies were initiated in the late 19th century to examine changes in soils after native grassland was converted to agricultural crops. In the Missouri study, the agronomist J.W. Sanborn tested the sustainability of soil-nutrient supply and crop production under several management systems (Brown 1994), and Sanborn and his successors have used these studies to recommend cropping systems to promote high crop yields and soil fertility. In the Oklahoma

Table 4.2. *A selection of long-term soil studies that demonstrate the global interest in quantifying the sustainability of soils and ecosystems. It is significant that nearly all of these studies are conducted on agricultural soils with relatively high native fertility*

Research site	Location	Soil Taxonomy order[a]	Crop description	Date of origin	Reference
Park Grass	Rothamsted, UK	Alfisol (Paleudalf)	Grass cut for hay	1856	Thurston et al. 1976; Tilman et al. 1994
Sanborn Field	Columbia, MO, USA	Alfisol (Ochraqualf)	Corn, wheat, crop rotations	1888	Brown 1994
Askov Experiment	Askov Experimental Station, Denmark	Inceptisol (Ochrept); Alfisol (Hapludalf)	Crop rotations	1893	Schjonning et al. 1994; Christensen et al. 1994
Old Rotation	Alabama, USA	Ultisol (Udult)	Cotton and legumes	1896	Mitchell et al. 1996
Bad Lauchstädt	Bad Lauchstädt, Germany		Various crops	1902	Körschens 1994
Bretton Plots	Bretton, Alberta, Canada	Alfisol (Boralfs)	Wheat–legume rotation	1930	McGill et al. 1986
Arlington Plots	Madison, WI, USA	Mollisol (Argiudoll)	Continuous corn	1958	Vanotii and Bundy 1995
Tamworth Rotation	Tamworth, NSW, Australia	Vertisol	Legume–cereal rotation	1966	Holford 1981
Haryana	Hisar, India	Inceptisol (Ustochrept)	Millet–wheat rotation	1967	Gupta et al. 1992
Coweeta Hydrologic Laboratory	Otto, NC, USA	Ultisol (Hapludult)	Oak–hickory forest	1970	Knoepp and Swank 1994
Yurimaguas	Yurimaguas, Peru	Ultisol (Paleudult)	Corn–bean rotations	1972	Smyth and Cassel 1995

[a]*Soil Taxonomy* Suborder or Great Group in parentheses (Soil Survey Staff 1998).

study, only a few years after converting native prairie grassland to wheat, phosphorus became deficient, yet when phosphorus was added in fertilizer, soil nitrogen continued to be mineralized to provide highly productive wheat yields without any inputs of fertilizer nitrogen *for 65 years* (Webb et al. 1980). These latter studies help illustrate the incredibly high native fertility of many Mollisol soils (Table 4.1).

For good reason, long-term studies of agronomic soils have now been initiated throughout the world to help answer questions similar to those posed by Lawes, Gilbert, and Sanborn about the sustainability of soil and plant productivity in agro-ecosystems (Table 4.2). Long-term studies of agricultural soils now include corn in Yurimaguas, Peru, and Illinois, USA; wheat in central India, Australia, and in Oklahoma and Oregon of the USA; rice in the Philippines; and cotton in Alabama, USA. These soil-ecosystem experiments test soil effects of practices such as crop rotations, cultivation, manuring, weed control, and fertilizers. A recent survey of long-term field studies of agro-ecosystems in the USA indicated that there were four studies > 100 years in age, 12 studies that were > 50 to 100 years, and 25 studies > 25 to 50 years (Mitchell *et al.* 1991).

A CRITIQUE OF THE SCIENCE OF SOIL AND ECOSYSTEM CHANGE

At the beginning of the 21st century, soil and ecological sciences are in a remarkable period of development. Highly informative ecological data that describe the functioning of soils and ecosystems have proliferated. Significant scientific studies on soil-ecosystem processes are incredible in number (e.g., Stone and Kszystyniak 1977; Troedsson 1980; Ulrich *et al.* 1980; Alban 1982; Vogt *et al.* 1983; Sánchez *et al.* 1985; Binkley 1986; Pastor and Post 1986; Swank and Crossley 1988; Driscoll *et al.* 1989; Federer *et al.* 1989; Jenkinson 1991; Nowak *et al.* 1991; Ugolini and Sletten 1991; Johnson and Lindberg 1992; Leigh and Johnston 1994; Binkley and Giardina 1997; Schlesinger 1997). Ecological theory and computer simulation that integrate soil and ecosystem processes have grown rapidly as well (e.g., Shugart 1984; Cosby *et al.* 1985; O'Neill *et al.* 1986; Reuss and Johnson 1986; Huston 1994; Ågren and Bosatta 1996). Though there may be much to be critical about in this large body of science (e.g., Stone 1975; Hurlbert 1984; Peters 1991), in aggregate these efforts represent an incredible progress of science over a relatively short period of a few decades.

Despite this scientific progress, we have serious criticisms about the current state of these ecological sciences, especially about the ability with which these sciences address soil and ecosystem change through time. There are at least five reasons why we are not well prepared to move rapidly beyond our current understanding of soil and ecosystem change.

First, nearly all long-term soil-ecosystem studies are confined to agricultural ecosystems (Table 4.2). There are few long-term studies of soil change in grasslands, pastures, and forests (Powers and Van Cleve 1991; Pregitzer and Palik 1997; Vance 1998). These systems function differently than annual agro-ecosystems and, ironically, our understanding of soil change in long-lived systems such as forests is based almost entirely on short-term observation.

Second, nearly all long-term soil-ecosystem studies are conducted on relatively fertile soils. Most long-term studies currently underway are based on Mollisols, Alfisols, and Vertisols (Tables 4.1, 4.2), which not only have relatively high pH but also can have substantial native soil fertility. Few long-term studies investigate advanced weathering-stage soils, although such soils cover extensive areas of the biosphere (Buol *et al.* 1989; Richter and Babbar 1991). Since even fertile soils can be greatly altered by soil management, long-term studies are clearly needed on less fertile soils that have more limited abilities to supply nutrients at a rate that keeps pace with plant demand.

Third, many long-term soil-ecosystem studies (Table 4.2) are not experiments at all. Nearly all studies more than 50 years old lack any experimental design or independent plot replication. This unfortunate situation is due to the fact that the oldest studies pre-date the development of modern statistical methods, which are products of the early to mid-20th century (Fisher 1951). Although few scientists would reject the oldest soil studies out of hand ("as merely anecdotes"), it is important that results from the classical studies be carefully interpreted, due to their deficient experimental designs. Like it or not, most of the oldest studies are carefully tended observations of "$n = 1$."

Fourth, long-term agricultural studies focus mainly on yield response or soil change rather than on the processes that affect the soil and plant response (Anderson 1991). With notable exceptions, changes in yield or soil properties have been the chief response variables of interest in long-term studies, with much less attention being given to the processes that affect response. Given the rationale and the value of the studies that are already underway (Table 4.2), long-term experiments that examine both response *and* processes are greatly needed to better understand sustainability of managed soils and ecosystems.

Fifth, long-term ecological studies are almost entirely focused on a relatively narrow set of natural science issues that involve current land management regimes and ecological processes. Only in recent years have historical and ecological analyses of landscapes, ecosystems, and soils begun to interact (Cronon 1983; Christensen 1989; Newman 1997; Foster 1999).

5

Soil change over millennia, centuries, and decades

In this book, we examine soil change and the processes affecting soil change over time scales of millennia, centuries, and decades. We use a particular region, the southeastern United States, and indeed a particular farm, the Old Ray Place in South Carolina, to examine soil change over this sweep of time. To evaluate system change, standard methods are used to analyze physical and chemical properties of soil, water, plant, and rock (Environmental Protection Agency 1971; Page 1982; Klute 1986; Carter 1993).

The three time scales correspond to three generic ecosystems that dominated large expanses of southeastern North America over the last few millennia, centuries, and decades: the upland, mainly oak–hickory deciduous primary forest; the cultivated agro-ecosystem mainly aimed at cotton; and the secondary post-agricultural pine forest, respectively (Table 5.1). Changes in soil that have occurred through this sweep of time have been affected by inputs, transformations, translocations, and losses of materials and energy. Table 5.1 summarizes input and output processes in the three ecosystems over the three time scales.

Advanced weathering-stage soils predominate on upland landforms across southeastern North America. Soils are typically ancient Ultisols (Table 4.1), and in the southern Piedmont, where the Old Ray Place is located, soils have been subject to intense weathering well into the Tertiary. At the time that upland forests began to be cleared and converted to row-crop agriculture, many of these soils had relatively low concentrations of organic carbon and nitrogen, and low levels of bio-available forms of phosphorus, calcium, magnesium, and potassium.

Following forest clearing, agricultural use has had distinctive effects on soil fertility. Forest clearing and burning were initially accompanied by a major input of nutrients. Cotton, tobacco, corn, and other crops accelerated nutrient losses from increased erosion,

Table 5.1. *Three time scales of ecological processes that affect soil change in much of the potentially arable uplands of southeastern North America*

Time scale	Generic ecosystem	Dates	Processes controlling system change		
			Input	Output	
Millennial	Primary deciduous forest	Pre-18th century	Atmospheric deposition Mineral weathering Nitrogen fixation	Hydrologic leaching Fire Denitrification Erosion	
Centurial	Agriculture for cotton, corn, and wheat	18th to mid-20th century	Atmospheric deposition Mineral weathering Nitrogen fixation Fertilization/manure amendments	Accelerated erosion Hydrologic leaching Harvest removals Fire Denitrification Accelerated decomposition of soil organic matter	
Decadal	Secondary pine forest	Mid-20th century to present	Atmospheric deposition Mineral weathering Nitrogen fixation Soil organic matter reaccumulation Potential fertilizer amendments	Hydrologic leaching Potential harvest removals Potential fires	

decomposition of organic matter, hydrologic leaching, and crop harvests. Cropping also brought substantial nutrient enrichments from fertilization and liming, especially after the late 19th and early 20th centuries.

After 100 to 250 years of agricultural use, cultivated fields across the region were abandoned in great numbers, and have succeeded to secondary pine forests, most of which no longer receive fertilizer or lime inputs. A major effect of reforestation is the physical stabilization of soils, many of which had been seriously eroding under agriculture (Trimble 1974; Richter *et al.* 1995a). On the other hand, forests have grown rapidly and have exerted an intense demand for soil nutrients. How these soils, previously disturbed by agriculture, are now responding to forest regrowth has implications for many advanced weathering-stage soils throughout the world that are undergoing substantial and not dissimilar land-use conversions.

The three parts to the book allow us to place soil-chemical change of the current ecosystem in the context of soil and ecosystem change affected by the system's past land use and the system's longer-term pedogenesis as well. This is by definition a cumulative ecological analysis as it examines not only current processes affecting change but also the legacy of past inputs, transformations, translocations, and losses.

SOIL CHANGE OVER MILLENNIA: EFFECTS OF LONG-TERM PEDOGENESIS

To evaluate the most significant soil processes that form advanced weathering-stage soils, we examine in Part II of this book the bio-geochemistry of solid, solution, and gas phases of deep soil profiles, specifically in the Piedmont of southeastern North America. Results of these investigations describe not only the extremely weathered state of these soils, but also the biogenic origin of acidification and mineral weathering.

The southeastern USA is a good region to consider problems associated with soil change and sustainability. Most soils in the region are acidic Inceptisols and Ultisols (Table 4.1), soils with relatively low native fertility, composed of materials that have long been subject to intense mineral weathering, acidification, and leaching loss. Many soils of this humid region, including those at the Calhoun Experimental Forest, are residual, derived directly from the bedrock below them. These Ultisol soils have had many or most of their primary minerals consumed by weathering and are composed mainly of secondary minerals that are resistant to further dissolution by weathering. Secondary minerals include low-activity clays such as kaolinite and hydrous oxides of iron and aluminum. Soils with many of these characteristics occupy two to three billion hectares world-wide, mainly in the warm temperate zone and across the tropics (Table 5.2).

A complex of ecological processes in warm, humid climates push soils toward acidification and an advanced stage of weathering (Table 1.1). Soil acidification results mainly from the biogenic formation of organic and mineral acids followed by hydrologic leaching of reaction products, so that over time losses of chemical elements outpace inputs and re-supply (Table 5.1). A battery of biogenic acids combines to weather soils to such extreme states. These include organic acids in surface layers and in rhizospheres of deeper substrata; sulfuric acid from microbial oxidation of sulfides; nitric acid from microbial oxidation of mineral N; and carbonic acid from belowground respiration of roots and microbes. If soils are geomorphically stable (i.e., if they do not erode, are not disturbed by glaciation, or are not buried or transported by wind-blown loess or water-borne fluvial deposits), biogeochemical weathering can be so intense that soils (O through C horizons) develop that are many meters deep (Buol and Weed 1991; Stolt *et al.* 1992; Richter and Markewitz 1995b).

The pedogenic processes that form acidic, highly weathered soils suggest why these soils require careful management to sustain their supply of nutrients. Because advanced weathering-stage soils such as Ultisols, Oxisols, and related acidic soils are so common world-wide, our results from these soils of southeastern North America may provide insights about soil change and sustainability of soil-nutrient supply for a large area of the earth's landscape (Table 5.2).

SOIL CHANGE OVER CENTURIES: LEGACIES OF PAST LAND-USE ON SOIL

"Agriculture was the South," writes environmental historian A.E. Cow-drey (1996) about agriculture as the dominating influence on regional ecology, economy, and culture in the 19th century. In fact, agricultural use of soil in southeastern North America was substantial as far back as the turn of the first millennium, a time in which Native Americans had domesticated many plant species for cultivation. Throughout the Mississippian Period of about 800 to 1500 AD, prodigious quantities of these food crops were harvested in the region (Larson 1980; Dobyns 1983; Asch and Asch 1985a, 1985b; Cowdrey 1996). For good reasons, however, Native Americans in the southeast based their agriculture not on the

Table 5.2. *Areal distribution of Ultisols and Oxisols in the tropics (FAO–UNESCO (1974, 1988) data from the* Soil Map of the World, *with computations in Richter and Babbar 1991). Units are in millions of hectares or in tens of thousands of square kilometers. The data are for soils in the FAO classification that are of the Acrisol and Ferralsol orders, closely similar to Ultisol and Oxisol, respectively. These data may well underestimate areas in South America and Africa covered by Ultisols and overestimate those of Oxisols (Richter and Babbar 1991)*

Region	Ultisols (millions of hectares)	Oxisols (millions of hectares)
Total tropics	547	1048
Central America[a]	21	1
South America[b]	190	614
Africa[c]	85	418
Asia[d]	251	15

[a]FAO–UNESCO (1975).
[b]FAO–UNESCO (1971).
[c]FAO–UNESCO (1977a).
[d]FAO–UNESCO (1977b, 1979).

highly weathered soils of the uplands but almost exclusively on riparian soils of stream and river terraces (Hudson 1976). The alluvial soils had high native fertility relative to nutrient removals in crop harvests and, in the end, effects of their management of soils were relatively minor.

By the late 1600s, however, a new colonial agriculture had spread through the Atlantic Coastal Plain of the southeastern USA. By the 1680s and 1690s, large areas of the low country forest of Virginia and the Carolinas had been converted to a new agriculture (Figure 5.1). European- and African-American populations grew rapidly (Figure 5.2) as Native American populations diminished (Wood 1989; Thornton *et al.* 1992). Land grants were issued, first adjacent to the port cities of Charleston and Wilmington in the Carolinas, then along river bottoms throughout the Atlantic Coastal Plain, and by the mid-1700s up into the interior Piedmont. As human populations grew, the largely deciduous forests of the uplands were cleared, and the region's upland Ultisols converted to agricultural use. Trees were both felled and girdled, and forest biomass not removed by harvest remained on site to be burned and to decompose (Williams 1989; Whitney 1994). Fire was used to speed decomposition of branches and stems, and to provide plant nutrients to the otherwise nutrient-depauperate Ultisols of the new agricultural fields.

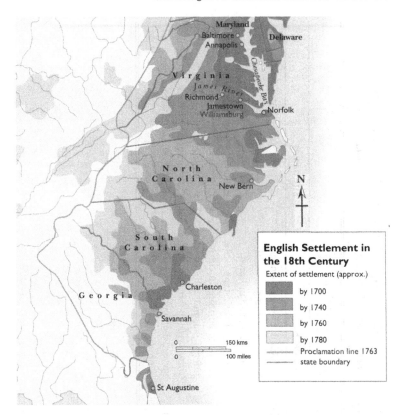

Figure 5.1. Map illustrating the early expansion of European and African populations in eastern North America from 1600 to 1810 (Homberger 1995).

By the turn of the 19th century, forests were being converted to farms throughout Virginia, the Carolinas, and Georgia. In the early decades of the 1800s, farming for cotton, tobacco, corn, and wheat rapidly moved westward, through Alabama, Mississippi, Louisiana, and into Arkansas and Tennessee (Cochrane 1979). This manner of agriculture was problematic, not only because the agricultural system was based on a soil resource largely of the Ultisol order (Table 4.1), but also because so much of the power to plow, harrow, weed, and harvest was supplied by human slaves.

Such agricultural development initiated a process that has substantially transformed many soils of the region, physically, chemically, and biologically. Prior to the USA's Civil War in the 1860s, agricultural fields were mainly derived from newly cleared forest, from "fresh soil." After the soil's productive capacity began to decline, land was

Table 5.3. *Summary of the soil-comparative study conducted in the vicinity of the Calhoun Experimental Forest, SC. All sites were selected on interfluves derived from granitic bedrock with similar slope class (<10%) and soil series. Table 13.1 summarizes additional ecological information*

Ecosystem	Land-use history		Site names
	Approx. 1800–1940s	Post-1940s	
Old hardwood: white oak, hickory	Deciduous forest: grazed woodlot; uncultivated, unfertilized, and unlimed	Deciduous forest: no evidence of cutting or burning; uncultivated, unfertilized, and unlimed	Calhoun, Murphy, Mt. Zion, Rt. 196
Old-field loblolly pine	Cotton and other row crops: cultivated, fertilized, and limed	Loblolly pine forest: no evidence of cutting or burning; uncultivated, unfertilized, and unlimed	Calhoun, Murphy, Padgett's Creek Church, Rt. 196, Greer
Hayfield agriculture	Cotton and other row crops: cultivated, fertilized, and limed	Grass hayfield: fertilized, limed, and harvested of hay several times a summer	Calhoun, Murphy, Padgett's Creek Church, Mt. Zion, Greer

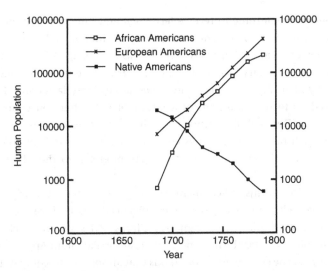

Figure 5.2. Populations of Native Americans, Europeans, and Africans, 1665 to 1790 in the Carolina Piedmont and Coastal Plain (Wood 1989).

abandoned. After several years of a forest fallow, the old fields may have been brought back into cultivation by re-clearing and burning, which at least temporarily regenerated nutrient availability.

After the Civil War, however, tenant farming, sharecropping, and a cash economy emerged throughout the region. Cotton continued to expand as a dominant agricultural crop. Fields became more permanent as agricultural practices were modified to include fertilizers, lime, and manure. Fertilizer and lime amendments expanded greatly in the late 19th and early 20th centuries, as did the use of new crop varieties. On a few fields, crop rotations with leguminous cowpeas and clover helped add organic matter and nitrogen to the soil. Although slavery had been abolished, the human condition all across the region was extremely difficult, to say the least.

In Part III of this book we evaluate the legacy of this past agricultural land use, by examining a number of modern Ultisol soils that have experienced distinctly different land-use histories. Four uncultivated and unfertilized soils that remain under hardwood forests are compared with a number of soils that were cultivated as early as 1800, the time when many upland forests were initially cleared for cotton (Table 5.3). The cultivated soils in this soil-comparative study have undergone one of two land-use trajectories. One land-use trajectory resulted in soil being long cultivated for cotton and other crops beginning as early as 1800, and which by the mid-20th century were converted to hayfields. These five old fields continue to receive fertilizers and lime to maintain hay production. The second land-use trajectory resulted in soil being culti-vated for cotton and other crops beginning as early as 1800, but which by the mid-20th century were abandoned of agriculture and converted to forest by plantings of loblolly pine (*Pinus taeda*). These latter sites have received no further additions of fertilizer or lime after conversion to pine forest. Detailed observations have been made of the upper 1 m of soil in each of the three ecosystems (Tables 5.1, 5.3), and differences between the soils are attributed to long-term effects of their land-use histories.

All sites in this soil-comparative study are located on nearly level interfluves with <10% slopes, and are advanced weathering-stage soils derived from similar granitic-gneiss bedrock. All soils are of closely similar soil series, for the most part the common Piedmont Appling and Cecil series. The critical assumption was that all soils in the comparative study were closely similar prior to about 1800, whether they currently support hayfields, old-field pine stands, or hardwood forests.

Results of the comparative study demonstrate how substantially soils have been altered by past agricultural use. In short, agricultural practices have depleted soils of some chemical elements such as carbon, but also greatly enriched these soils in nitrogen, phosphorus, and cal-cium. The pH of soils has been greatly elevated by former liming practi-ces, with exchangeable acidity reduced at least through the upper 1 m of mineral soil. The legacy of agriculture appears to persist for many decades and may even continue for centuries after agricultural aban-donment and reforestation.

SOIL CHANGE OVER DECADES: SOIL EFFECTS OF CONTEMPORARY ECOSYSTEMS

In the 20th century, the southeastern region of the USA, long dominated by agriculture, entered an era of more diverse land uses. Large areas of previously cultivated land were abandoned, especially in the Piedmont. Naturally regenerated and planted pine forests (loblolly, shortleaf, slash, and longleaf) now cover about 25 million hectares of the southern USA (Powell et al. 1993), almost all on previously cultivated soils. The region also contains >10 million hectares of mixed oak–pine forests, much of which has also been previously cultivated (Richter and Markewitz 1995a).

Throughout the 20th century, these old-field pine and oak-pine forests grew at rapid rates. These forests are also harvested for large

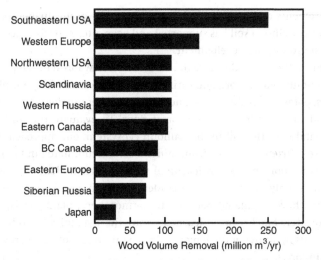

Figure 5.3. More wood is currently being harvested for industrial products from the forests of the southeastern USA than from any other wood-producing region in the world (Jaakko Pöyry Group 1994). In the southeastern USA, effectively all wood production is from secondary forests, most of which grow on previously cultivated soils.

volumes of wood, with removals in the early 1990s totaling about 250 million m^3 per year (Powell *et al.* 1993). Remarkably, more industrial-wood fiber is currently grown and harvested in the forests of the southeastern USA than in any other wood-producing region in the world (Figure 5.3). In large measure, such harvests are derived from the same soil resource that was so extensively used by agriculture in the past. How the nutrient requirements of such rapidly regrowing southern forests are met by soil-nutrient supplies is not yet well documented and this is an important part of this book's objective.

The ability of soil to sustain its nutrient supply while meeting nutrient requirements of modern ecosystems depends not only on the current system's nutrient demands, but also on how soil-nutrient supply is controlled by millennial-scale pedogenic processes and past land uses. To examine soil changes affected by contemporary ecosystems, a replicated experiment is used that is located at the Calhoun Experimental Forest in South Carolina. In the winter of 1956–1957, pine seedlings were planted on permanent resampling plots that were laid out in old-cotton fields last cultivated in 1954. Trees have been remeasured periodically during the subsequent four decades of forest development and soils have been resampled seven times during this interval, with nearly all soil samples permanently archived at Duke University. The repeated forest measurements and archived soil samples provide an invaluable record of how the Calhoun soil has supported and been altered by four decades of forest growth and development.

From these four-decade data, we estimate how forest development has accumulated soil organic carbon, and transferred nitrogen, phosphorus, potassium, calcium, magnesium, and other nutrients from mineral soil to forest biomass and forest floor. To evaluate the nutrient sustainability of the soil in the Calhoun ecosystem, one approach is to compare nutrient removals from mineral soil (accumulated in biomass and forest floor and lost to leaching) with long-term changes in soil-nutrient supply. Over the four decades, relatively large quantities of nutrients have circulated between the atmosphere, plants, soils, and groundwater. Much can be learned about the dynamics of soils and ecosystems by studying the long-term nutrient cycling of the regrowing Calhoun forest.

In the next chapter, ecological details of the Calhoun ecosystem are described. Specific topics include its climate, geology, vegetation, paleoecology, geomorphology, watershed hydrology, land-use history of the Old Ray farm, and the relationship of the Calhoun ecosystem to a broader landscape. Although the specific topic is the soil of the Old Ray farm, also very much at issue is change in soils and ecosystems throughout the temperate zone and the tropics.

6

The Calhoun forest: a window to understanding soil change

The Calhoun Experimental Forest was established by the United States Forest Service in the Sumter National Forest as a principal area for research on forest stabilization of eroding agricultural soils and for the management of secondary southern pine forests. The land that became the Sumter National Forest was consolidated in the 1920s through the 1940s almost entirely from old Piedmont farms.

LOCATION OF THE CALHOUN ECOSYSTEM

The Calhoun Experimental Forest is located at about 34.5° N, 82° W, about 20 km southwest of the small town of Union, South Carolina, the county seat of Union County (Mabry 1981; Charles 1987). Union lies about 450 km southwest of Richmond, Virginia and 300 km northeast of Atlanta, Georgia. From the Calhoun ecosystem, the 2000 m peaks of the Great Smoky Mountains are about 100 km to the northwest, and the beaches of the Atlantic Ocean about 280 km to the southeast.

The Calhoun is part of the southern Piedmont physiographic province, an ancient region that is about 100 to 350 m above sea level. The region lies between the Atlantic and Gulf Coastal Plain to the east and south, and the Appalachian Mountains to the west and north.

GEOLOGY

The geologic material underlying the Calhoun ecosystem is partly metamorphosed granitic gneiss, a most common geologic material in the southern Piedmont (Figure 6.1). The material is siliceous and coarse in texture. When weathered to soil, almost all of the primary minerals (such as biotite and feldspars) are lost by weathering dissolution

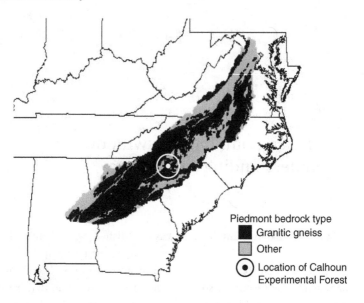

Figure 6.1. Granitic gneiss is the most common bedrock in the southern Piedmont region of eastern North America.

throughout the upper meters of the soil. Total elemental analyses of these profiles are summarized in Table 6.1, which shows that elements such as Ca, Mg, and Na are greatly depleted from the upper soil layers due to long-term weathering and hydrologic removal. Minor chemical elements in the profile are summarized in Appendix IV.

Pre-metamorphic emplacement of the ignaceous granite that later became the Piedmont occurred in the Cambrian or late Precambrian, >500 million years BP. The granite experienced slight metamorphism near the end of the Paleozoic, about 300 My BP, and was uplifted through the early Mesozoic, about 200 My BP. Various sedimentary deposits covered the Piedmont for much of this time, and in the southern Piedmont, from what would become central Virginia through eastern Alabama, sedimentary layers were removed by erosion in the early Cenozoic, about 70 My BP.

Geologists hypothesize that a steady state of chemical weathering, erosion, and isostatic uplift has operated in the Piedmont since the early Cenozoic (Hack 1960; Pavich 1985). Apparently, isostatic uplift has compensated physical and chemical denudation and the Piedmont's surface has been relatively stable over a very long span of time. Most remarkable

Table 6.1. *Total element concentrations for soils from the Calhoun Experimental Forest, SC. Soil profiles were collected from the four experimental blocks of the long-term field study between 1990 and 1994 (site P-1). Rock samples were collected from the bottom of a gully system along nearby Sparks Creek*

Depth (m)	Horizon	Si	Al	Fe	Ti	K	Ca	Mg	Na
					$(mg\ g^{-1})$				
0–0.15	A + E	432.5	13.1	7.14	3.047	4.915	0.232	0.164	0.211
0.15–0.35	E	418.7	27.0	10.67	3.975	5.618	0.241	0.375	0.331
0.35–0.6	EB	356.5	86.6	28.22	4.895	5.467	0.408	1.023	0.328
0.6–1.0	B	305.1	129.5	44.51	5.253	4.978	0.296	1.249	0.083
1.0–1.5	B	304.4	127.7	42.78	4.886	6.455	0.144	1.114	0.111
2.0–2.5	B	315.4	125.6	34.57	4.514	12.22	0.113	1.773	0.399
3.0–3.5	BC	312.8	126.3	31.25	4.015	21.29	0.666	3.833	2.089
4.0–4.5	CB	330.6	100.6	24.41	3.546	39.68	0.719	3.388	3.338
5.0–5.5	C	332.2	98.0	24.06	3.365	36.71	1.966	3.939	6.658
6.0–7.0	C	329.6	98.9	19.64	2.490	27.46	8.789	5.079	12.83
7.0–8.0	C	326.4	99.7	19.48	2.706	30.18	8.247	5.238	14.76
Rock	R	324.9	88.1	20.88	2.069	9.791	27.23	5.276	36.21
					Coefficients of variation (%)				
0–0.15		2	33	18	15	6	19	27	47
0.15–0.35		1	43	35	15	4	3	31	17
0.35–0.6		9	32	23	0	17	10	9	58
0.6–1.0		18	40	30	5	2	49	7	51
1.0–1.5		13	26	18	13	38	32	29	2
2.0–2.5		5	15	2	1	23	27	28	8
3.0–3.5		0	2	2	12	55	114	4	23
4.0–4.5		2	13	7	13	14	132	7	85
5.0–5.5		4	14	32	31	3	126	22	104
6.0–7.0		4	6	43	30	26	137	23	127
7.0–8.0		2	16	27	8	26	133	5	124
Rock[a]		0	0	1	0	0	3	2	0

[a]Rock densities of 2.52 g cm^{-3} indicate little evidence of chemical weathering.

perhaps is to consider that the rate of isostatic uplift may be controlled in part by the rate of biogeochemical weathering that involves mineral weathering, soil formation, and hydrochemical removal.

PALEOECOLOGY

The southern Piedmont of North America has experienced a range of climatic and vegetative change since the early Tertiary (over about

70 My), the time over which the region has also been physiographically stable (Pavich 1985). Forests have covered this landscape for most of this long sweep of time, although temperatures have fluctuated between boreal and tropical (O'Neill 1985: Wolfe 1985; Delcourt *et al.* 1993).

In the Paleocene (about 55 to 65 My BP), a period of great evolution and expansion of angiosperm species, relatively warm, humid conditions promoted closed-canopy, tropical rain forests across much of southeastern North America. Through much of this period temperature across the region ranged from 20 to 30 °C.

During the Eocene (about 40 to 55 My BP), climates in southeastern North America were more varied. For the most part, average temperatures remained much warmer than today (20 to 30 °C). Periodic episodes of cooling occurred throughout the Eocene (Wolfe 1985), apparently in response to large volcanic eruptions that affected much of the global climate. Rainfall was strongly seasonal, and vegetation ranged from tropical rain forests to seasonally dry tropical and subtropical forests.

During the Oligocene, Miocene, and Pliocene (35 to 2 My BP), climatic fluctuations became increasingly severe. Extinctions affected forests such as those in southeastern North America, changes accompanied by wide temperature fluctuations. The tropical forests in southeastern North America waned as warm temperate and cool temperate deciduous forests became predominant in the region. To the west of the Piedmont in the Appalachian Mountains, cool temperate and boreal conifers such as spruce (*Picea*) and hemlock (*Tsuga*) extensively covered higher elevations (Delcourt *et al.* 1993).

Climatic fluctuations continued during the Quaternary (2 My to present). More than 20 glaciation episodes have characterized this period in earth's history, although southeastern North America has not been directly affected by glacial ice (van Donk 1976). The composition of plant species in southeastern forests has responded accordingly (Delcourt *et al.* 1993). The mass animal extinctions during this latter period of geologic history have no doubt had an impact on soils (Kurten and Anderson 1980). Giant ground sloths and mastodons, and as many as two-thirds of North America's mammalian genera, were lost near the end of the Pleistocene, about 10 000 years ago (Haynes 1991).

RECENT CLIMATE AND WATERSHED

The current warm temperate climate of the Calhoun ecosystem is classified as humid continental, with long, hot summers and short, mild winters. Annual precipitation averages about 1250 mm (1973–1987,

Whitmire, South Carolina), annual evapotranspiration about 850 mm, and annual drainage loss (surface runoff plus groundwater recharge) approximately 400 mm (Gnau 1992). Precipitation is relatively constant throughout the year, whereas evapotranspiration is highest in summer and drainage losses highest in winter and early spring. The soil moisture regime is classified as udic (rarely droughty for more than 90 days). Air temperature averages about 16 °C, ranging in most years between about −5° and 40 °C. Under forest cover, soil temperature at 0.2 m depth ranges between about 5 and 25 °C. The soil has a thermic temperature regime with soil temperatures that average about 16 °C.

The Calhoun ecosystem and nearly all of the southern Piedmont drain directly to North America's Atlantic coastline. Runoff from the Calhoun ecosystem drains first to Sparks Creek and Padgett's Creek, two small tributaries of the Tyger River. After about 30 km, the Tyger flows into the Broad River, which then joins the Congaree River and finally the Santee River. The Santee empties into the Atlantic Ocean about 70 km northeast of Charleston, South Carolina. During the 20th century, Lakes Moultry and Marion were constructed by the US Army Corp of Engineers to divert a fraction of the Congaree–Santee water to the Charleston Harbor via the Cooper River. Past agricultural use of the Piedmont has transported enormous amounts of sediment into these two reservoirs (Trimble 1974; Patterson and Cooney 1986; Richter *et al.* 1995a).

RECENT VEGETATION

In the late 20th century, rural uplands of the southern Piedmont are mainly covered by three vegetation types: grassland managed for hay and pasture; old-field pine or oak–pine forests that grow on formerly cultivated soils; and remnant, mixed oak–hickory hardwood forest, which across most of the Piedmont covers steep slopes, rocky soils, and plastic clays.

Prior to about 1750, the Calhoun area supported primarily a deciduous forest. Common species of this forest were white, black, southern red, and northern red oaks (*Quercus alba, Q. velutina, Q. falcata,* and *Q. rubra*); shagbark, mockernut, and pignut hickories (*Carya ovata, C. tomentosa,* and *C. glabra*); loblolly, shortleaf, and Virginia pines (*Pinus taeda, P. echinata,* and *P. virginiana*); and many other species such as yellow poplar (*Liriodendron tulipifera*), sweetgum (*Liquidambar styraciflua*), red cedar (*Juniperus virginiana*), sourwood (*Oxydendron arboreum*), dogwood (*Cornus florida*), red maple (*Acer rubra*), black gum (*Nyssa sylvatica*), redbud (*Cercis canadensis*), persimmon (*Diospyros virginiana*), hornbeam

(*Ostryra virginiana*), and ironwood (*Carpinus caroliniana*). Such species are described well in Radford *et al.* (1968) and Harlow *et al.* (1979).

As the Carolina Coastal Plain and Piedmont became subdivided by the colonial land grants in the 17th and 18th centuries, the upland, mixed hardwood forests were cleared and much of the landscape converted to agricultural fields. Since then, most soils in the Piedmont, including those at Calhoun, have been used to support crops of cotton (*Gossypium* spp.), corn (*Zea mays*), tobacco (*Nicotiana tabacum*), wheat (*Triticum aestivum*), as well as woodland and open pasture for pigs and cattle.

In the summer of 1954, the last crop of cotton on the Calhoun research area was harvested and was followed by a two-summer fallow. In the winter of 1956–1957, loblolly pine seedlings were planted, beginning four decades of forest and soil studies (Zahner 1967; Wells and Jorgensen 1975; Harms and Lloyd 1981; DeBell *et al.* 1989; Binkley *et al.* 1989; Buford 1991; Richter *et al.* 1994; Markewitz *et al.* 1998).

SOILS AND GEOMORPHOLOGY

Soil of the upland Calhoun ecosystem is characterized by its advanced weathering stage, lack of primary minerals, and its prominent acidity and textural gradient with soil depth. The soil is classified as an Ultisol (Table 4.1, Figure 6.2), one of the world's most common soils, that typically reside on relatively stable landforms and predominate in southeastern North America (Schoeneberger 1995), much as they do in many other warm temperate and tropical regions throughout the world (Table 5.2).

In addition to geomorphic stability, leaching is a prerequisite for Ultisol soils. Ultisols are formally defined by their acidity and elevated clay in B horizons. They are closely similar to Acrisols in the FAO–UNESCO classification, Red-Yellow Podzols in the 1938 USDA system, *Podzolicos Vermelho Amarelo Distroficos* in the Brazilian system, and *Sols ferralitiques fortement et moyennement desatures* in the French ORSTOM system (Buol *et al.* 1989; Richter and Babbar 1991).

The specific Ultisols at the Calhoun ecosystem are classified as a clayey, kaolinitic, thermic Typic Kanhapludult of the Appling series. The Appling series is an extremely acidic soil derived from granitic-gneiss bedrock. Appling soils are closely related to those of the Cecil series, the most common soil series in the southern Piedmont. Together, the Cecil and Appling soils cover approximately a third of the southern Piedmont (Figure 6.3).

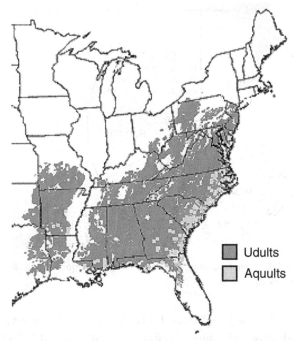

Figure 6.2. Map of Ultisol suborders in eastern North America illustrating the predominance of Ultisol soils. Udults are primarily well drained upland Ultisols (with udic moisture regime), whereas Aquults are Ultisols with high water tables (with aquic moisture regime).

Appling soils are distinguished by their relatively thick, coarse, surficial horizons which are sandy loams or loamy sands. These surficial layers are the A and E horizons, that vary in thickness from 0 to >0.6 m depending on past agricultural erosion. The surficial A and E horizons overlie about 2 m of slightly mottled, acidic, clayey B horizons that are dominated by kaolinite clay minerals which are coated with precipitates of hydrous Fe and Al oxides. Below the B horizons are >5 m of acidic, highly weathered C horizons, known as saprolite. At the Calhoun pine forests in the 1990s, rooting has been directly observed to extend to 4 m depth and probably extends more deeply.

THE PIEDMONT FARMLAND THAT BECAME THE CALHOUN ECOSYSTEM EXPERIMENT

The experimental ecosystem at the Calhoun forest is located on land originally granted by King George III and the State of South Carolina to Daniel Prince (in 1769, 1773, and 1774), John Prince (in 1786), Richard

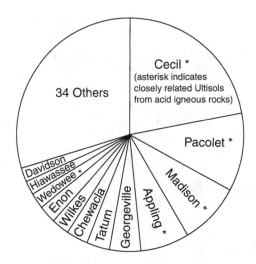

Figure 6.3. Relative frequency of areal coverage of major soil series in the southern Piedmont, USA.

Halcome (in 1772), and John Barnett (in 1774). The royal grants reserved for the British crown all white pine (*Pinus strobus*) trees and 10% of all silver and gold that were found on the property. Neither white pines nor gold were found, although the Daniel Prince and Richard Halcome farms included many acres of fertile alluvium along the Tyger River to the northeast (Figure 6.4). Such bottomland may well have been previously planted by Native Americans in corn and other crops, and a large Native American village of the Mississippian tradition (800 to 1500 AD) has been identified about 5 km downstream of the old Prince and Halcome farms on a fertile terrace of the Tyger River (J. Bates, personal communication, USDA Forest Service, Sumter National Forest, Whitmire, SC).

From the 1820s to the 1840s, Rev. Thomas Ray (b. 1780, d. 1862) bought and consolidated the Prince and Halcome farms and combined them with adjacent properties to create a large 1835-acre plantation (about 745 ha). The farm became known as the Old Ray Place, a name that endured into the 20th century. From 1806 to 1861, Rev. Ray was the ordained minister of the still-active Padgett's Creek Baptist Church, on whose grounds the extended Ray family is buried (Sparks 1967). With his wife, Sarah Whitlock, Rev. Ray raised a large family on the farm that produced high yields prior to the Civil War (Table 6.2), when cotton was profitable and the household and farm economy were supported by slaves. Ray was noted for serving many small, nearby churches, including the locally famous Baptist Church at Fairforest (Owens 1971; Spears

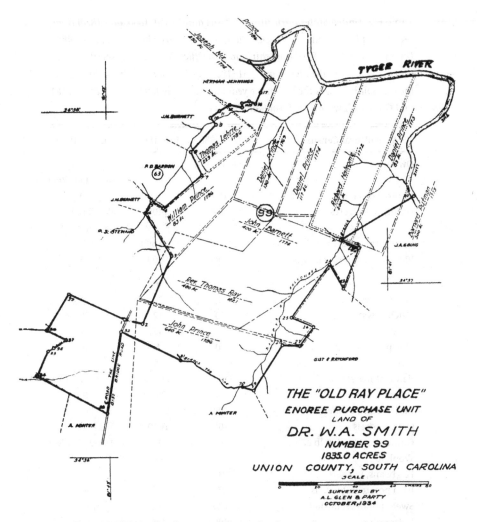

THE "OLD RAY PLACE"
ENOREE PURCHASE UNIT
LAND OF
DR. W.A. SMITH
NUMBER 99
1835.0 ACRES
UNION   COUNTY, SOUTH CAROLINA
SCALE
SURVEYED BY
A.L. GLEN & PARTY
OCTOBER, 1934

Figure 6.4. Map of land grants of farms that became known as the Old Ray Place, now a part of the Calhoun Experimental Forest, SC. This map was made in 1935 for the US government's court case that seized the Old Ray Place from Dr. W.A. Smith.

1974). Although the ownership of the Ray farm would change hands several times after the preacher's death, the farm boundaries established by him remained intact until the land was sold to the US government in the 1930s to become Enoree Purchase Unit 99 of the Sumter National Forest.

After Rev. Ray's death in 1862, the farm was assumed by Ray's son, Jack, who was not able to run the farm successfully in the years

Table 6.2. *United States Agricultural Census data (1850, 1860, and 1870) from the farm of Rev. Thomas Ray. Census data are from Schedule 4, South Carolina, Union County, Cross Keys Township and Post Office. Material prosperity marked the period prior to the Civil War (1861–1865), typical of agriculture throughout the Old South, as was the loss in the years immediately following. (Sites P-1, H-1, and G-1 are all within the farm boundaries.)*

US Agricultural Census	July 1850	July 1860	August 1870
Owner	Rev. Th. Ray	Rev. Th. Ray	Jack Ray (son of Rev. Ray)
Improved acres	400	600	60
Unimproved acres	1436	1400	Woodland 100 Other unimp 1700
Cash value of farm	$16 524	$32 000	$8000
Value of farm implements	$382	$600	$250
Horses	9	16	3
Mules/asses	8	10	3
Milch cows	20	12	7
Other cattle	29	45	14
Sheep	30	1	0
Swine	75	90	20
Value of live stock	$1775	$4500	$700
Wheat, bushels	175	365	0
Indian corn, bushels	2750	3000	200
Oats, bushels	400	1000	0
Ginned cotton, 400 lb bales	86	98	10 (450 lb bales)
Wool, lb	70	—	—
Pea/beans, bushels	10	500	—
Irish potatoes, bushels	15	200	—
Sweet potatoes, bushels	100	600	—
Butter, lb	300	400	—
Barley, bushels	—	25	—
Value of homemade manufactures	$64	$73	—
Value of animals slaughtered	$400	$895	$100
Value of all farm products including betterments	NA[a]	NA	$1100

[a]NA, data not accumulated in Census.

immediately following the Civil War (Table 6.2). In 1876, the 1835-acre farm was sold to James Long at public auction for $6050. Until the turn of the 20th century, Long apparently farmed the land through tenant farmers and sharecroppers. Union County Court Affidavits from 1935 indicate that Long raised cotton, corn, winter wheat, rye, oats, cattle, and pigs, and maintained fences, built outbuildings, and cut firewood and timber. After Long's death in 1899, debts from Long's estate caused the farm to again be sold at auction, this time to Dr. W.A. Smith in 1901 for $1800. Union County Court Affidavits from 1935 indicate that Dr. Smith managed the farm for three decades much like Long had before him.

During the international economic depression of the early 1930s, Dr. Smith did not pay property taxes on the Old Ray Place, which totaled about $150 per year. In 1935, he sold the farm to the federal government for $6422, or $3.50 per acre. The land became part of the newly forming Sumter National Forest, a National Forest particularly notable for its many agriculturally eroded soils and gully systems.

After the land was sold to the federal government, the fields on which the Calhoun soil-ecosystem experiment reside were continued in cotton for two decades via tenant farmers. The last crop of cotton was harvested in the fall of 1954. Aerial photographs illustrate that in the 1930s, the 1940s, and early 1950s the specific fields that would become the Calhoun Forest Experiment continued to be cleanly cultivated (Figure 6.5).

RELATIONSHIP OF THE CALHOUN ECOSYSTEM TO SOUTHEASTERN USA ECOSYSTEMS

The Calhoun Forest Experiment warrants our interest not only because it is a unique, long-term soil-ecosystem experiment, but also because the site has many similarities with a much larger region. The Calhoun ecosystem's geologic substrata, current vegetation, soil conditions, and land-use history are all closely similar to conditions across much of the southeastern North America.

The Calhoun's geologic material is granitic gneiss, bedrock that underlies well over half of the southern Piedmont (Figure 6.1). The dominant soil series is the Appling series, a close relative of the Cecil soil and other Ultisols that occur commonly in the southeastern USA region (Figures 6.2). The Cecil soil covers more than 20% of the southern Piedmont: the Appling soil covers somewhat less than 10% (Figure 6.3).

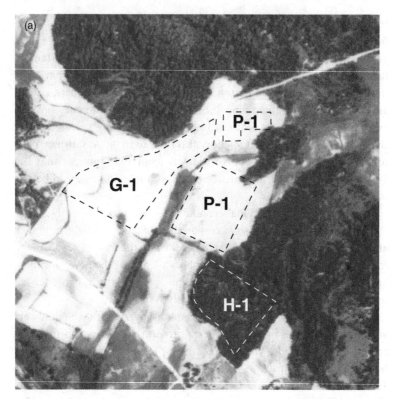

Figure 6.5. Aerial photographs of the Calhoun Forest Experiment, SC taken (a) in 1944, prior to the beginning of the study, (b) in 1970, and (c) in 1994. Cleanly cultivated cotton is illustrated in the 1944 photograph when it extends across future hayfields, G-1, and the future pine forest, P-1. In 1970 and 1994, hayfields are evident in G-1, and the experimental pine stands of P-1 are (b) 13 and (c) 37 years in age. The uncultivated hardwood forest, H-1, is illustrated in all three photographs.

The patchwork of the old fields at the Calhoun Experimental Forest (Figure 6.5) indicates a history of erosive land use that is shared across much of the southeastern USA. On the experimental plots themselves, erosion losses due to agriculture range between 0 and 0.25 m. Erosion-control terraces, installed during the early soil stabilization programs of the USDA's Soil Conservation Service (now the Natural Resources Conservation Service), cut on contour across several of the now forested permanent plots.

The Calhoun loblolly pine ecosystem is closely similar to many planted and naturally regenerated pine stands that grow on previously cultivated land in the southeastern USA. In the 1990s, the area of

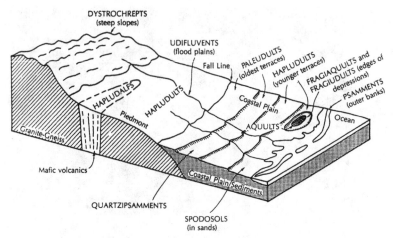

Figure 6.6. Landscape diagram of Ultisols and other related soils in a portion of the Carolinas (Buol *et al.* 1989).

southern pine ecosystems across 13 southeastern states in the USA totaled about 25 million hectares.

THE CALHOUN ECOSYSTEM'S RELATIONSHIP TO LANDSCAPES WORLD-WIDE

When scientists of the US Forest Service selected the Calhoun Experimental Forest for an experiment on pine growth in the southeastern United States, they chose well. A number of ecological characteristics of the Calhoun ecosystem are similar not only to those across the southeastern USA but also to a much larger landscape across the warm temperate zone and the tropics as well (Sánchez 1976; Richter and Babbar 1991; Jordan 1998).

Advanced weathering-stage soils such as Ultisols and Oxisols (Table 1.1) are very common soils world-wide (Table 5.2) and are soils that commonly occur on the most stable landforms in the soil landscape (Figures 6.6, 6.7). According to the FAO *Soil Map of the World*, Ultisols cover more than 250 million hectares of Asia, including southeastern China, the Shan Plateau in Burma, the Indochinese and Malaysian peninsulas, the Philippines, and non-volcanic areas of the Sumatran and Kalimantan islands in Indonesia (FAO–UNESCO 1979). Ultisols cover enormous areas of South America including western Amazonia, the upper Rio Madeira, and Araguaia basins. In Africa, they cover the west African basement complex and the Inter-Rift Valley of Tanzania.

Moreover, many tropical forests supported by Ultisols and Oxisols

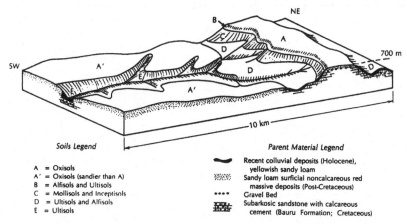

Figure 6.7. Landscape diagram of Oxisols, Ultisols, and other soils in the vicinity of Echapora in São Paulo State in Brazil (Lepsch *et al.* 1977). The advanced weathering-stage soils, Oxisols and Ultisols, occupy landforms that are physically most stable.

have been, and are currently being, cutover and converted to pastures and a variety of agricultural uses. An enormous tropical landscape is being subjected to land-use conversions that are not dissimilar to those that have already affected the Calhoun and much of southeastern North America.

We are confident in suggesting that changes in the Ultisol soils of the Calhoun ecosystem and of southeastern North America have implications for soil and ecosystem dynamics across a vast and important area of the earth's surface.

Part II

Soil change over time scales of millennia: long-term pedogenesis

Natural soil formation is often celebrated for producing fertile soils. Good examples are organic-rich, pH-neutral soils that support productive grasslands and river bottom forests. However, on physically stable surfaces of the uplands, humid-zone ecosystems affect a soil formation that is an acidifying and often nutrient-depleting process. Though soils are not often completely exhausted of their energy and nutrients, soils with low native fertility cover several billion hectares of the world's 13 billion hectares of soils. When converted to human use, such potentially depauperate soils are disposed to fertility problems unless managed with care.

7

Soil development from the Devonian to Mendocino and Hawaii

One of the most extraordinary outcomes of biological evolution has been the co-evolution of soil and forest ecosystems. Soil and forest ecosystems have been major components of the biosphere for nearly 400 million years, since the Devonian period of the earth's history. While many biological species have long gone extinct, soil and forest systems have managed to sustain themselves over this enormous period, a time that has included widely fluctuating environments, cataclysmic geological events, and massive biological extinctions.

Most remarkable are soils and terrestrial plant life from the Devonian period (Gensel and Andrews 1984; Algeo *et al.* 1995), when for the first time forests and soil expanded as a bio-mantle over the earth's continental surface (Figure 7.1). These significant terrestrial ecosystems are evidenced in fossils, not only in the form of tree trunks, foliage, roots, and animals, but also in the form of their underlying fossilized soils, known as paleosols (Retallack 1990, 1992).

Prior to the Devonian, in the Silurian period, plant and animal life was already advanced, but continental surfaces were not densely vegetated nor deeply rooted. They were covered by proto-soil and bare rock. Bacteria, fungi, algae, and lichens were active in pre-Devonian soils (Wright 1985), but across continental surfaces there were no well rooted plants to anchor surficial material in place. Prior to the stability provided by Devonian forests, unconsolidated materials of the earth's continental surface were readily disturbed and eroded by wind, ice, rain, and runoff.

During the Devonian, trees created complex forest ecosystems. Trees grew with large diameter, great height, and deep roots (Beerbower 1985). The *Callixylon* or *Archaeopteris* trees, for example, were more than

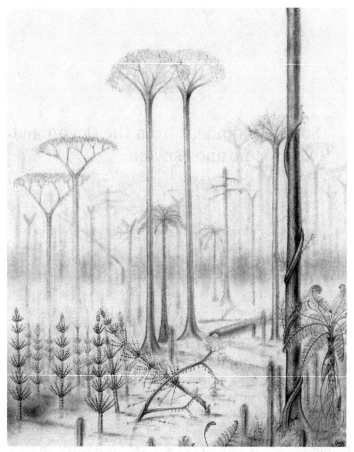

Figure 7.1. Devonian forests were complex, complete with producers, consumers, and symbionts (Behrensmeyer *et al.* 1992).

1 m in diameter, and have been found in Devonian deposits of North
America and Europe. Many trees were spore-forming tree-ferns
(Pteridophytes or Progymnosperms), others were seed-producing
Gymnosperms. Nearly all are now extinct. The Sphenophytes, for
example, were once a large, diverse group of trees, vines, and shrubs, but
are now limited to a small number of diminutive horsetail species
(*Equisetum*).

In these ancient forest ecosystems, the detritus of tree biomass
was decomposed and recycled by a diverse community of soil bacteria,
fungi, and macrofauna. The Devonian forests had high-level consumers,
including insects, earthworms, and spiders. Soil-borne fungi symbioti-
cally associated with plant roots to form mycorrhizae, the fossil remains
of which clearly demonstrate the complexity of Devonian forests.

Devonian forests transformed the soil environment, biologically, chemically, and physically. The gas, liquid, and solid phases of Devonian forest soil developed a strong imprint of biological activity. Like forests today, forest roots, soil macrofauna, and microbial communities greatly increased the surface area of the biota–mineral contact.

Under humid climates and in freely draining materials, Devonian root and microbial action expanded the depth of soil development. In addition to the physical mixing of soils caused by plants and animals (known as pedoturbation), respiration of belowground biota elevated CO_2 in soil atmospheres, and carbonic acid in soil solutions stimulated weathering of minerals. Organic acids and microbial oxidation of sulfur to sulfuric acid also accelerated acidification and weathering of the soil's solid phase. As the earth's crust came under increasing biotic control, surficial soil erosion diminished, and a diversity of soils were able to form.

SOIL GENESIS CAN ENRICH AND DEPLETE SOILS OF NUTRIENTS

Natural soil formation is often celebrated for creating fertile, organic-rich soil. Forest and grassland ecosystems benefit soil fertility by accumulating organic matter and conserving nutrients within the ecosystem. Nitrogen also is accumulated by biological N_2 fixation. But whether soil supports forest or grassland, *soil is an open system, entirely subject to material gains or losses.*

Soil-comparative studies along the Mendocino coast of California (Jenny *et al.* 1969) and on the Hawaiian islands of Hawaii, Molokai, and Kauai (Chadwick *et al.* 1999) are used here to help initiate our perspective of soil genesis in southeastern North America and in much of the humid temperate zone and tropics as well.

HANS JENNY'S MENDOCINO STAIRCASE

Along the exquisitely beautiful northern California coastline in Mendocino County, a series of five marine terraces has been uplifted about 30, 50, 90, 130, and 200 m above current sea level during the last million years (through about half the Quaternary period). Beach deposits that cover all five terraces are derived from the same graywacke sandstone, and Jenny (1980) refers to the set of uplifted terraces as the Mendocino Staircase (Figure 7.2). The area currently has a Mediterranean climate with wet, snow-free winters and cool, dry summers. Rainfall averages

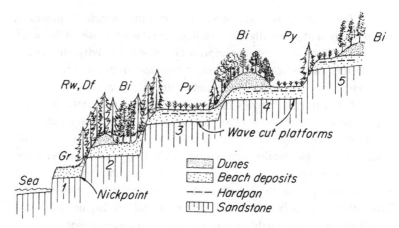

Figure 7.2. Hans Jenny's (1980) diagram of the Mendocino chronosequence. West to east sequence of elevated marine terraces covered with beach material and sand dunes along the northern California Pacific Coast. Gr, grassland; Rw, redwood; Df, Douglas fir; Bi, Bishop pine; and Py, Pygmy forest.

970 mm and mean temperature is about 12 °C. Jenny assumed that all soil-forming factors "save age t" have been comparable during soil formation.

The Mendocino chronosequence of level beach deposits has seasonally high water tables that accumulate on the hard sandstone that is several meters below ground surface (Figure 7.2). On the youngest beach deposits, grasses, lilies, peas, buttercups, and sunflowers quickly establish on the new soil surface. The prolific grasses and herbs increase organic matter in the soil-rooting zone, and legumes enrich soil organic matter with nitrogen derived from symbiotic N_2 fixation. Thus is formed an Entisol soil (Table 4.1) with its surficial A horizon enriched in organic matter. As organic matter accumulates and deepens the A horizon under grasses and herbs, the youthful Entisols are transformed into Mollisol soils (Table 4.1). Mollisols are typically associated with grassland ecosystems; they have especially deep A horizons that are dark, organic rich, relatively high in pH, and highly fertile. This surface layer is distinctive and is classified as a mollic epipedon (a deep, high-pH, organic-enriched A horizon).

On the second terrace, forests are well established, with redwood (*Sequoia sempervirens*), bishop pine (*Pinus muricata*), and Mendocino cypress (*Cupressus pygmaea*). Weathering proceeds rapidly in these soils under forest, as weatherable minerals such as potassium-containing feldspars are weathered and lost from surface soils (Figure 7.3). The

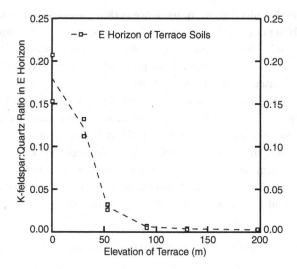

Figure 7.3. In the soils of the Mendocino Staircase, K-feldspar is rapidly decomposed in surficial layers of soil forming from graywacke beach deposits. Terrace elevations reflect soil age, and data plotted are ratios of sand-sized (0.05 to 2 mm) K-feldspar to quartz (Gardner 1967). Since the quartz is highly resistant to dissolution, it provides a reference to gauge the disappearance of feldspar through time. The most youthful soils forming at sea level are Entisols, while soils at 30 m are Entisols and Mollisols, at 53 m are Alfisols and Ultisols, and at >91 m are groundwater Spodosols.

Mollisols are transformed into Alfisols and their more acidic relatives, the Ultisols (Tables 1.1, 4.1). The early stage of relatively high soil fertility is not long-lasting in these Mendocino ecosystems.

On the third, fourth, and fifth terraces, soils are transformed into Spodosols (Table 4.1). Soils develop an iron-pan, an indurated layer cemented by iron hydroxides at 50 to 75 cm depth. The iron-pan further restricts vertical drainage and water tables are commonly perched within the upper 75 cm of the soil's surface. These soils develop extreme acidity, and are so affected by potentially toxic aluminum and poor drainage that the few plants that do survive can barely grow. Lichens (*Cladonia* and *Usnea* spp.) are prolific in these environments. Tree species include the bishop and Bolander pines (*Pinus contorta* var. *Bolanderi*), trees that grow to the diminutive height of only 1 to 3 m.

The Mendocino chronosequence illustrates how ecosystem development leads initially to a transient stage of relatively high fertility and neutral pH: Inceptisol, Mollisol, and Alfisol soils that supported high biotic productivity of dense grasslands and tall forests. What is most striking in the Mendocino chronosequence is that long-term soil development leads to plant stress and substantial reductions in plant productivity. Although the Mendocino trajectory is an extreme example of soil and ecosystem change, the pygmy forest provides a powerful symbol for the potential consequences of a soil genesis.

Hans Jenny never tired of talking about chronosequences on field trips to the Mendocino Staircase. Many visitors were impressed by these visits, including the agriculturist Wes Jackson (1992), who with eloquence described how a field trip with Jenny fundamentally shook his ideas about ecology, ecosystems, and soil. Confronted with the pygmy forest, Jackson realized that widely held ideas about nature and ecosystem development need much more critical examination.

SOIL DEVELOPMENT IN THE HAWAIIAN ARCHIPELAGO

Because soils are open systems, they are rarely completely exhausted of their nutrients and energy. A soil chronosequence with wide application to large areas of Ultisols and Oxisols found throughout the world (Table 5.2) is one currently being studied in the Hawaiian archipelago (Vitousek *et al.* 1994; Chadwick *et al.* 1999). The chronosequence of six soils is located on three Hawaiian Islands: Hawaii, Molokai, and Kauai. Soils are derived from basaltic lava and lava–ash mixtures which range in age from 300 years to 4.1 million years. They represent a dramatic sequence of soil development from volcanic Andisols to extremely weathered Oxisols (Table 4.1). All soils have similar elevation and climate and support natural ecosystems dominated by *Metrosideros polymorpha* trees and *Cibotium* tree-ferns. Crews *et al.* (1995) explicitly discuss assumptions of the soil comparisons.

Over millennial time scales, the Hawaiian soils (upper 1 m depth) exhibit profound changes in elemental composition, mineralogy, and bioavailability of nutrients (Crews *et al.* 1995; Torn *et al.* 1997; Vitousek and Farrington 1997; Chadwick *et al.* 1999). Soils that are most productive for plant growth are those intermediate in age (e.g., 20 000 to 150 000 years). Soil organic matter and its contents of carbon, nitrogen, and phosphorus increase from the beginning of soil genesis to intermediate soil ages and then decline at differing rates thereafter. This pattern of accretion and depletion is directly controlled by the course of clay-mineral weathering. During the first 150 000 years, volcanic materials weather to organophilic, amorphous, non-crystalline clays such as allophane, imogolite, and ferrihydrite. The surfaces of such clays are extensive, hydrated, and possess a variable charge that is well suited for binding organic matter strongly. The organo–clay bonds are strong enough to protect organic compounds from microbial decomposition. As a result, organic carbon builds up to $> 50\,kg\,m^{-2}$ in soils that are 20 000 to 150 000 years in age.

The non-crystalline clays are, however, unstable beyond 150 000 years, and clay mineralogy of older soils becomes dominated by crystalline minerals such as low-activity kaolinite, halloysite, gibbsite, goethite, and hematite. Surface areas of such clays are much reduced, as are their affinities for organic matter. Soil organic carbon, nitrogen, and phosphorus are much diminished beyond 150 000 years. Carbon in the upper 1 m depth, for example, decreases by about 50% between 150 000 and 1 million years. Carbon isotope data provide a rich set of details about the dynamics of soil organic matter over this sweep of time.

Like the Mendocino Staircase, the Hawaiian chronosequence illustrates how advanced weathering-stage soils are profoundly depleted of primary minerals and nutrient elements such as phosphorus and nutrient cations (Crews et al. 1995). As phosphorus in primary minerals is rapidly diminished (from 82% of total P at 300 years to 1% at 20 000 years), phosphorus is accumulated in organic matter. Bioavailable phosphorus increased during this intermediate period, as indexed by resin-extractable phosphorus. However, as Oxisols are formed, bioavailability of phosphorus is greatly diminished. Such decreases in phosphorus bioavailability are supported by various evidence: the accumulation of occluded soil phosphorus by iron oxides, the strong phosphorus immobilization in decomposition experiments, the positive plant response to phosphorus in fertilizer amendments, and low soil and plant indices of bioavailable phosphorus (Crews et al. 1995). Net primary productivity peaks in this chronosequence at 150 000 years, but remarkably is diminished by only about 15% by 4.1 million years, hardly a pygmy forest.

A CRITICAL ROLE FOR INPUTS

In Hawaii, the Oxisols currently experience significant phosphorus deficiency, and may well be deficient in nutrient cations as well. Relative to biotic demands, sources of these nutrients are few from within the ecosystem. The main potential source for nutrient input is not from weathering dissolution of soil minerals but from atmospheric deposition. In the Hawaiian Islands, cations are supplied by marine aerosols (especially enriched in magnesium in sea-salt aerosols), whereas phosphorus is derived from dust particulates from central Asia, a source that is more than 6000 km distant. Although these renewals are meager on a yearly basis, they are apparently sufficient to lessen the likelihood that these natural systems will lose substantial potential productivity and develop into pygmy forests.

Although the Hawaiian system may not develop into a pygmy forest, sweeping changes in biogeochemical controls have occurred over soil and ecosystem function. Early in the Hawaiian soil development, factors regulating nitrogen availability control plant productivity, whereas late in soil development (after Andisols have been altered to Oxisols), bioavailability of phosphorus controls ecosystem productivity. Other edaphic factors of significance limiting ecosystem productivity in Oxisols include nutrient-cation deficiencies, and toxicities due to aluminum, manganese, and other metals.

Soil and ecosystem developments that lead to acidification and depletion of chemical elements raise significant questions about natural and managed ecosystems and soil. For example:

- What controls and processes combine to so deplete soils of their nutrients and lead to such extreme acidification?
- How do advanced weathering-stage soils respond to management when harvests increase nutrient removals?
- What ecological conditions, biogeochemical processes, and management practices can ameliorate these millennial soil developments?
- Can biotic species and varieties be selected and managed to tolerate and even ameliorate the soil fertility of advanced weathering-stage soils?

In the next three chapters, we turn to the soils of southeastern North America. We first examine the prototypical advanced weathering-stage soil of the Calhoun ecosystem, a soil that has not been cultivated or fertilized and that still supports a hardwood forest which we assume has never been completely cleared for agriculture. We then discuss soil formation of this Ultisol in a hypothetical chronosequence, and lastly examine some of the ecological processes that affect its formation over time scales of millennia.

8

Genesis of advanced weathering-stage soils at the Calhoun ecosystem

On the granitic-gneiss bedrock of the southern Piedmont, soil and ecosystem development has proceeded along a trajectory that is broadly reminiscent of the development of advanced weathering-stage soils in Hawaii (Chadwick *et al.* 1999). Many physically stable soils in humid, warm regions develop into advanced weathering-stage Ultisols (Figure 8.1). Physically stable soils in humid climates acidify and are depleted of primary minerals because they are open systems entirely dependent on input–output budgets of chemical elements and energy (Chadwick *et al.* 1990).

From the time that granitic gneiss is exposed to the atmosphere and vegetation establishes itself, organic-enriched layers of soil begin to accumulate in O and A horizons. In humid, warm climates such as those in southeastern North America, exposed C horizons accumulate organic matter in O and A horizons and develop within a few years into a simple Entisol soil (Figure 8.1). Although O and A horizons may accumulate relatively rapidly, the formation of a well developed B horizon will take much longer. Nearly all soils that support plants and other soil biota have surficial A horizons enriched in organic matter. The content of soil organic matter is controlled mainly by soil texture (the proportions of sand, silt, and clay), clay mineralogy (kaolin, montmorillonite, or allophane), and moisture status.

As even a hint of a B horizon develops, however, the young Entisol is transformed into an Inceptisol (Figure 8.1). The incipient B horizon, also known as a cambic B horizon, is a distinguishing feature of the Inceptisol. The Inceptisol's B horizon has been leached of relatively soluble compounds such as soluble salts and calcium carbonate, and is often slightly stained with iron oxides. A cambic B horizon which is identifiable only by its color is classified as a Bw horizon. Though iron oxides may only begin to stain the B horizon, the inception of an

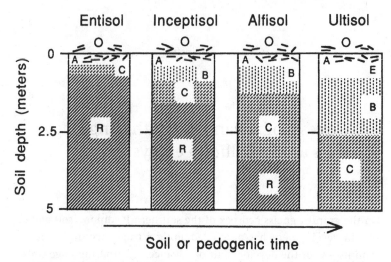

Figure 8.1. Soil formation in southeastern North America originating from granitic gneiss located on a physically stable interfluve surface (Richter and Markewitz 1995b).

O–A–B–C profile represents a stage of a larger set of soil-formation processes associated with illuviation, the B-horizon accumulation of hydrous Fe and Al oxides and clays.

On relatively level, stable landforms, B horizons accumulate clay as it is translocated (eluviated) from the A and E horizons and as it forms as a product of weathering and pedogenesis (Duchaufour 1982; Buol *et al.* 1989). As clay continues to accumulate in the B horizon, the aging Inceptisol is eventually transformed into an Alfisol (Figure 8.1). Although Alfisols are more weathered than the Inceptisols from which they were derived, they are relatively neutral in reaction and high in base saturation (the fraction of a soil's negative charges that is balanced by nutrient cations such as calcium, magnesium, or potassium). By definition, an Alfisol's clay-enriched B horizon has a base saturation >35%. Alfisols are often relatively fertile soils due to moderate pH and their primary minerals that still release nutrients upon weathering. Many Alfisols have ample water storage capacity due to fine-textured B horizons, a property of great significance for vegetation in climates with extended dry seasons. Alfisols are highly significant to agriculture, and are some of the world's most well studied soils (Rust 1983).

In humid climates, Alfisols eventually acidify to Ultisols mainly because leaching losses outpace nutrient inputs by mineral decomposition and atmospheric deposition for long periods of time. Ultisols are

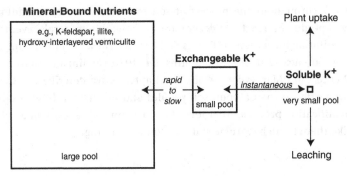

Figure 8.2. Mineral weathering is conceived as a set of dynamic chemical reactions that buffer nutrient removals. Removals of nutrient ions by root or microbial uptake, or from hydrologic leaching, tend to be rapidly buffered by nutrient ions that are exchangeable, and more slowly resupplied by those that are mineral bound. In advanced weathering-stage soils, weatherable minerals may be little able to buffer removals from the soil system.

generally similar to Alfisols in physical properties, but contrast greatly in their chemistry and nutrient status. In Ultisols, leaching has been so extreme that most if not all primary minerals are decomposed. Ultisols typically have meager release rates of calcium, magnesium, potassium, and phosphorus from mineral-weathering dissolution, rates that may be much lower than those in other soils in comparable environments.

In soils with weatherable primary minerals in the profile, the removal of weathering products from the soil solution, whether by root uptake or by hydrologic leaching, does not directly or immediately deplete bioavailable calcium, magnesium, potassium, or phosphorus, but rather helps ensure continued dissolution of minerals and supply of bioavailable nutrients (Figure 8.2). The significance of how removals stimulate release is not widely appreciated even though it often underlies the ability of soils to sustain their bioavailable supply of nutrients over millennial time scales. On the other hand, the rate of weathering dissolution is mineral dependent, and eventually primary minerals can be depleted by long-continued leaching. The trajectory of soil formation that leads to Ultisols and Oxisols is one in which a wide variety of weatherable minerals will be consumed. Thus, Schwertmann and Herbillon (1992) characterize advanced weathering-stage Ultisols and Oxisols as having "monotonous mineralogy," kaolin and oxide dominated. Such minerals are composed mainly of oxygen, hydrogen, silicon, aluminum, and iron and are incapable of maintaining a continuous supply of nutrients in soil solution.

Throughout the process of soil formation on stable landforms, acid-producing reactions decompose weatherable minerals, eventually facilitating formation of Ultisols and Oxisols. These acid-producing reactions are nearly all biogenic (Sollins *et al.* 1980; Ugolini and Slettin 1991; Richter and Markewitz 1995b). In Chapter 9, we describe the specific properties of the Calhoun soil profiles and in Chapter 10 we examine acidification processes that force the Calhoun soils and so many others like them to such extreme states of soil weathering.

9

The Calhoun soil profile

In southeastern North America, the last several centuries of land use have greatly altered nearly all of the region's soils. Yet, to understand natural soil formation over millennial time scales, we need to observe at least a few soils that have not been altered greatly by agriculture and related human disturbance.

Within a few kilometers of the Calhoun experimental ecosystem, we identified four Ultisol soils that support old forest ecosystems that appear to have been neither cultivated nor fertilized. All four soils are on nearly level, geomorphically stable interfluves, and are derived from granitic gneiss. Oak trees on these sites are relatively old (e.g., > 100 years). The four stands have no pines in the overstory or understory canopy, suggesting that these sites have not been completely cleared or greatly disturbed for many years. Although these hardwood forests were almost certainly used for timber, fuelwood, and free-range grazing, their soils have without doubt been disturbed far less than those that were converted for cotton, corn, wheat, and hay.

It is not easy to locate potentially arable soils in the Piedmont that have not been cultivated. Nearly all potentially arable soils have long ago been cultivated, and nearly all Piedmont hardwoods are on steep slopes, or are rocky or have heavy plastic clays. By carefully exploring the Piedmont hardwood forests, however, a few low slope, potentially arable soils can be found. In the next chapters, we describe these potentially arable soils that still remain under hardwood cover.

Some investigators have described such uncultivated Piedmont soils as "undisturbed" and even covered by "virgin forest" (e.g., McCracken et al. 1989). Such nomenclature is not altogether accurate, for although these soils may not have been cultivated, nearly all hardwood ecosystems were grazed by livestock and periodically harvested of timber and fuelwood throughout the 19th and early 20th centuries. Nonetheless, although the vegetation has been altered by past human

Table 9.1. Soil chemistry of the upper 5 m soil profile under uncultivated and unfertilized oak–hickory hardwood forest adjacent to the Calhoun Experimental Forest (site H-1)[a]. Soil series is Appling derived from granitic gneiss. The soil taxon is a Typic Kanhapludult

Horizon	Depth (m)	Organic C (%)	Organic N (%)	Ext P (μg g^{-1})	pH$_w$	pH$_s$	Exchangeable					ECEC	EBS (%)
							Ca	Mg	K	Na	Acidity		
							(mmol$_c$ kg^{-1})						
O1	—	51.4	1.22	—	—	—	158.0	97.9	31.4	0.4	33.4	321	89.6
O2	—	37.1	1.28	—	—	—	154.7	57.1	18.2	0.4	16.0	246	93.5
A and E	0–0.3	0.95	0.04	2.42	4.86	4.04	0.20	0.3	0.6	0.0	11.6	13.5	13.6
E	0.3–0.5	0.31	0.02	1.03	4.74	4.00	0.05	0.2	0.6	0.0	14.2	15.6	8.3
EB	0.5–0.75	0.21	0.02	0.33	4.75	3.80	0.10	2.2	1.4	0.0	27.8	31.9	11.7
B	0.75–1.0	0.12	0.01	0.12	4.81	3.86	0.10	2.9	1.0	0.0	37.5	42.8	10.7
B	1.0–1.5	0.090	0.009	0.00	4.57	3.96	0.19	1.2	0.8	0.1	43.5	45.7	5.0
B	1.5–2.0	0.058	0.004	0.08	4.69	4.00	0.04	0.6	0.4	0.2	33.1	34.2	3.3
BC	2.0–3.0	0.057	0.004	0.41	4.58	4.02	0.06	0.3	0.4	0.2	36.2	37.1	2.6
CB	3.0–4.0	0.049	0.003	0.41	4.43	3.94	0.04	0.2	0.7	0.2	45.4	46.5	2.3
C	4.0–5.0	0.045	0	0.98	4.57	4.18	0.05	0.4	0.7	0.3	30.3	31.7	4.3

[a] Samples of mineral soil are composites of two individual 10 cm-diameter cores. pH$_w$ is soil pH in water, 1 part soil plus 2 parts water; pH$_s$ is soil pH in salt, 1 part soil plus 2 parts 0.01 M CaCl$_2$. Ext P is P extractable in Mehlich III solution. Exchangeable Ca, Mg, K, and Na were displaced by 1 M NH$_4$-acetate at pH 7. Exchangeable acidity is extractable in 1 M KCl. ECEC is effective cation exchange capacity, or the sum of the four exchangeable cations plus KCl-exchangeable acidity. EBS is the effective base saturation of ECEC. Although EBS appears to be relatively high in the O horizons of the forest floor, we assume CEC (at pH 8.2) to be about 3000 and 2200 mmol$_c$ kg^{-1} in the O1 and O2 horizons (based on 1500 mmol$_c$ kg^{-1} carbon), respectively, and the base saturation (BS) of total CEC to be about 10% in both horizons. Analytical methods follow those described in Carter (1993), Page (1982), and Westerman (1990).

uses of hardwood forests, the soil profiles have been relatively little disturbed compared with regional norms.

Our criteria for identifying potentially arable soils that have not been cultivated were relatively strict. The soils have <10% slopes, are derived from granitic gneiss, and are classified as Appling, Cecil, or Madison soil series. The soils' A horizons are coarse textured (usually sandy loam) and have no evidence of accelerated erosion. The soils currently support relatively old hardwood forests (*Quercus alba* is dominant) that have no obvious signs of recent disturbance such as cutting or fire.

Although we judge that these four soils under hardwoods have not been cultivated and fertilized, of course we do not know this for certain. At worst case, however, given the age of the hardwood trees and the absence of pines, these soils have not been disturbed for a very long time. We are therefore confident that these soils have not been cultivated, and that fertilizer and lime inputs have been minimal to non-existent.

SOIL HORIZONS AND PROFILES

Soil horizons are layers of similar biogeochemistry and have been studied as key features of soil since the 19th century. Following a Russian convention, the nomenclature of soil horizons is based on a system of letters, e.g., O, A, E, B, and C, that proceeds downward from the soil's upper surface. Russian pedologists and ecologists, led by V.V. Dokuschaev, first described soil horizons and profiles according to soil texture, color, organic matter content, acid–base chemistry, and physical structure (Sibirtzev 1914). Although concepts of horizons have been increasingly enriched, for example by growing appreciation of geomorphology (Milne 1935; Daniels and Nelson 1987; Paton *et al.* 1995), we still characterize soil horizons with many of the same properties used by 19th century Russians (Soil Survey Staff 1975, 1992).

Table 9.1 summarizes soil data from one of these hardwood forests, site H-1 (Figure 6.5). The data indicate that soils under hardwood forests are remarkably depauperate but that they have well developed soil horizons, including organic O horizons lying above the mineral soil that includes A, E, B, and C horizons. Figure 9.1 illustrates the pattern of organic carbon and clay in the Calhoun soil profile. The figure illustrates how these two primary soil components change continuously with depth, not discretely as the concept of horizon may suggest.

In our study of the Calhoun soil, we examine the properties and formation of the full O- through C-horizon profile on geomorphically

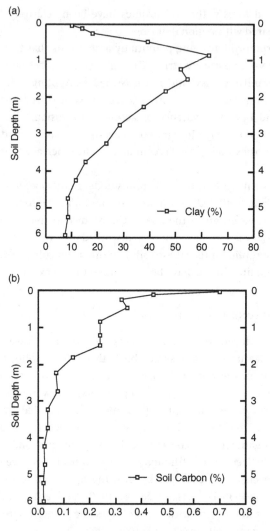

Figure 9.1. Soil organic clay ($<2\,\mu$m particles) (a) and carbon (b) concentrations as a function of soil depth in the Calhoun profile (site P-1).

stable interfluves that tend to allow formation of deep, advanced weathering-stage soils (Figure 9.2). In our perspective, soil represents the entire weathering zone of the earth's crust, and this entire volume is affected by pedological processes that are expressed by distinctive changes in soil chemical, physical, and biological properties.

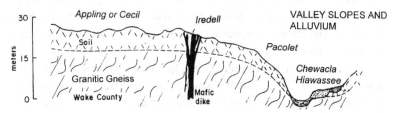

UPLANDS

INTERFLUVE

Figure 9.2. A cross-section diagram of a Piedmont soil landscape. Appling and Cecil soil series are typically located on the most level, geomorphically most stable interfluves.

THE CALHOUN SOIL PROFILE: THE APPLING–CECIL–MADISON SERIES

The following description of the Calhoun profile is derived from data assembled from four potentially arable but uncultivated Appling, Cecil, and Madison series soils that still support hardwood forest.

O horizons

The O horizons are derived mainly from oak and hickory foliage, wood, and bark, plus animal debris. Rates of decomposition, comminution (the bio-physical disintegration of materials), and biological incorporation into mineral horizons are all relatively rapid, and the mass of O horizons is typically less than $10\,\text{Mg ha}^{-1}$. The basal layers of these O-horizons have many active fine roots and soil macrofauna such as earthworms, insects, and myriapods that physically incorporate O horizon materials into the A horizon below. The O horizons are relatively acidic, with base saturation of the cation exchange capacity estimated to be about 10% (footnote in Table 9.1).

A horizons

Under organic-rich O horizons lie A horizons that are sandy loam or loamy sand layers that may be up to 30 cm in depth. Fine root biomass is most concentrated in surficial A horizons, and aggregation is most prominent in the uppermost layer of the A where soil organic matter is most concentrated. Structural aggregates, although present, are weak due to the coarse sandy texture. Bulk density is low, averaging as low as 0.8 Mg m^{-3} in the upper 0 to 7.5 cm layer of A horizons. Overall, the A and upper E horizons have modest accumulations of soil organic matter and nitrogen (0.95 and 0.041% organic carbon and nitrogen, respectively), and relatively low concentrations of bioavailable nutrients. In the A horizon, extractable phosphorus averages 2.4 µg g^{-1} and effective base saturation about 14% (Table 9.1).

The relatively coarse textures of surficial layers (sandy loam and loamy sand) limit the stabilization and accumulation of soil organic matter (SOM). Due to macroporosity and high diffusivity, these layers are well supplied with oxygen, and since the little amount of clay-sized minerals that is present (<15%) is low-activity kaolinite and quartz, there is little adsorption and stabilization of humified SOM (Figures 9.1, 9.3, 9.4). Within the Calhoun forest vicinity, A horizons with finer textures (loams to clay loams) accumulate 2- to 3-fold more SOM than is present in the sandy loam or loamy sand A horizons.

The Calhoun A horizons have high hydraulic conductivity and infiltration capacity, and A horizons retain relatively little water for plant uptake due to coarse sandy textures. Soil erosion is minimal under intact forest, and the main hydrologic export process is by deep drainage rather than by surface runoff.

E horizons

Below A horizons, the Calhoun soils have notably thick E horizons, sandy loam or loamy sand layers that have high hydraulic conductivity and little ability to store water. Surfaces of nearly all sand grains are uncoated by SOM, unlike the A horizon. In addition, clay, iron, and aluminum have been eluviated (translocated) from E horizons and have left not only a coarse texture, but the soil particles uncoated by hydrous iron and aluminum oxides (Table 9.1, Figure 9.1). The Calhoun E horizon is a remarkably simple horizon, dominated by coarse sand grains with low surface area. Lacking much SOM, the E has virtually no aggregation. Toward the basal layers of the E, however, a prominent transitional horizon is found, the EB, within which clays and hydrous iron and aluminum oxides increase as the B horizon is approached (Figure 9.1).

B horizons

As in many upland Piedmont soils, the B horizons at the Calhoun ecosystem are 2 to 3 m in thickness. Although the density of fine roots may be greater in more surficial layers, B horizons are extensively rooted. In fact, given the coarse texture of A and E horizons above, clayey B horizons are important for plant-water supply, especially during the middle and late growing seasons (e.g., June through October in the Carolina Piedmont). Despite the fact that only 20% of fine root biomass is located in B horizons, a hydrologic simulation (with the PROSPER model) suggested that more than 50% of annual transpiration was derived from water of the B horizon (Gnau 1992; Richter *et al.* 1994).

As is the case with more surficial layers, B horizons are extremely acidic and low in bioavailable nitrogen, phosphorus, calcium, magnesium, and potassium (Table 9.1). Although soil exchangeable calcium and magnesium may appear to be slightly elevated in subsoils, these are extremely low concentrations of these cations, especially for calcium (e.g., $>0.2\,\mathrm{mmol_c\,kg^{-1}}$ in the B horizons). The soil's inherent negative charge (its cation exchange capacity or CEC) is almost entirely balanced by acidic cations (Al^{3+} and H^+) throughout the entire profile. Acid saturation is highest (i.e., base saturation lowest) at $>1\,\mathrm{m}$ depths in lower B and C horizons (Table 9.1).

These B horizons are classified as kandic horizons, meaning they are very low in CEC per unit mass of clay. Kandic B horizons are composed of low-activity clay minerals, i.e., mainly kaolinite and gibbsite (Figure 9.3). Kandic horizons have an effective cation exchange capacity (ECEC) of $<120\,\mathrm{mmol_c\,kg^{-1}}$ of clay and a CEC of $<160\,\mathrm{mmol_c\,kg^{-1}}$ of clay (measured at pH 7).

The Calhoun B horizons are excellent examples of acidic, low-activity kandic subsoils that are characteristic of many Ultisols and even some Alfisols throughout the world (Richter and Babbar 1991). These kaolinitic materials are remarkably stable, and are relatively resistant to further mineralogic change, even under intense weathering conditions.

Although kaolin is clearly dominant, X-ray diffractograms suggest that hydroxy-interlayered vermiculite (HIV) is a secondary mineral in the Calhoun B horizons, probably a weathering product of mica. Occasionally, mica is prominent in the deeper layers of the Calhoun B horizons, at >2 to $3\,\mathrm{m}$ depth. Such mica may also be found in the underlying C horizons as well, and may thus buffer removals of bioavailable potassium.

C horizons

Below the kandic clay-enriched B horizon, the C horizons at the Calhoun are many meters deep: $>5\,\mathrm{m}$ in all hand augering and hydraulic probing at Calhoun (Figure 9.4). On some interfluves, bedrock may be as much as

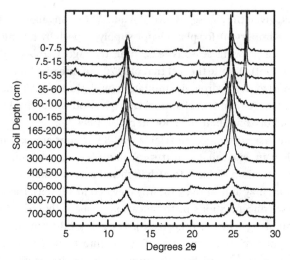

Figure 9.3. Mineralogy of Calhoun soil horizons (0 to 8 m depth) as illustrated by X-ray diffraction. The predominance of kaolinite (0.7 nm peak) in the profile is attributed to the Ultisol being an advanced weathering-stage soil. The Calhoun soil is formally classified to be in the Appling series which is a fine, kaolinitic, thermic Typic Kanhapludult. Site P-1.

20 to 30 m below the soil surface, although characteristics and depths of soil above the unweathered bedrock are not well quantified (Hunt 1986; Buol 1994; Creemans *et al.* 1994; Graham *et al.* 1994). The C horizons are certainly rooted but at lower root densities than B horizons. Bioavailable nitrogen, phosphorus, calcium, magnesium, and potassium are all extremely low in these layers with organic carbon <0.06% (Table 9.1). The C horizon's acidity is remarkably extreme (Table 9.1), with some layers having base saturation <3% of the soil's ECEC.

Sometimes in the deepest layers of the C horizon, directly on top of bedrock, is a soil layer undergoing active mineral weathering (Calvert *et al.* 1980). These layers of the C horizon may possess weatherable primary minerals such as biotite and feldspars and are denoted as Cr horizons. They were not sampled in the soil profile in Table 9.1 and are presumed to be >10 m.

THE ADVANCED WEATHERING STATE OF THE SOIL PROFILE

Below the C horizon is unweathered granitic-gneiss bedrock, which had a pH in water of 7.9 after it was pulverized. Since the pH of all soil sampled throughout many meters of A to C horizons ranges from 3.8 to

Figure 9.4. Sampling deep soil profiles can be accomplished by hand with a bucket auger to about a 6 m limit or by hydraulic probe which can sample soil solids, solutions, and gas samples from depths as great as 30 m. Illustrated is a truck-mounted Geoprobe which is well suited to investigate the full soil profile (photograph Michael Hofmockel).

4.2 in 0.01 M $CaCl_2$, the Calhoun soil is nothing less than extremely acidic in its native condition.

Exchangeable acidity (in 1 M KCl) totals about 3000 $kmol_c$ ha^{-1} in the 6 m profile, an enormous quantity of acidity. More impressive still is the quantity of acid that has been *consumed* during the weathering and transformation of the granitic-gneiss bedrock to the kaolinite-dominated Ultisol. Transforming 1 m of granitic gneiss into kaolin is estimated to require (i.e., to have consumed) on the order of 100 000 $kmol_c$ ha^{-1} of acidity (Richter and Markewitz 1995b). Weathering 10 m of granitic gneiss to kaolinite required about 10^6 $kmol_c$ ha^{-1} of acidity.

Pavich (1985, 1986) estimated that weathering Piedmont granite proceeds at a rate of 7 to 40 m per million years. A weathering rate of 10 m per million years would require about 1 $kmol_c$ ha^{-1} per year of acid input, a not unrealistic flux of acidity as we will see in subsequent chapters.

This extreme acidification raises significant questions about the sources of acidity that have so thoroughly weathered the Calhoun soil. In the next chapter, we examine sources of acidity that have so strongly altered these soil profiles through pedogenic time.

10

The forest's biogeochemical attack on soil minerals

90   In the southern Piedmont of North America, many soils are not only highly weathered, they are also often 5 to >30 m deep above unweathered rock. Typical is an upper 1 to 3 m of O, A, and B horizons, below which is the C horizon saprolite of highly variable depth.

This extreme result of weathering and soil genesis is not peculiar to southeastern North America. In peninsular Malaysia, about 20 m of weathered and highly acidic C horizon saprolite is derived from granite (Eswaran and Bin 1978). Other highly weathered profiles that are 10 to 100 m depth are found on the island of Hong Kong (Ruxton and Berry 1957; Carroll 1970), on the island of Sumbawa in Indonesia (D.D. Richter, unpublished data), and in the Amazon of Brazil (Nepsted et al. 1994).

ECOSYSTEM ACIDIFICATION OVER MILLENNIAL TIME SCALES

Deep, highly weathered soils are the products of soil reactions initiated by biogenic acids. These acid-producing reactions are important not only to soil genesis and to the weathering of the earth's crust, but also to the chemistry of natural waters, the atmosphere, and oceans (Chizhikov 1968; Jenny 1980; Holland 1984; Ugolini and Slettin 1991; Berner 1995; Richter and Markewitz 1995b; Jones and Mulholland 1998). This is why the Devonian period of the earth's history is so significant to the earth's biogeochemistry. The Devonian period brought the first forest ecosystems that spread across continental areas, forests that initiated profound effects not only on soils, but also on the weathering crust, natural waters, and the atmosphere itself.

Acidification processes are not well quantified, and five sources of acidity that lead eventually to advanced weathering-stage soils will be examined in this chapter (Table 10.1). The main sources of acidity

Table 10.1. *Five acid systems that drive soil genesis and crustal weathering*

Acid system	Internal ecosystem sources	External sources
Organic acids	Leaching of forest canopy, forest floor, roots, and decomposition products add organic acids to soil water (Qualls and Haines 1991). Especially significant in A horizons, upper B horizons, and rhizospheres (Cronan et al. 1978). These acids complex and mobilize metals	Atmosphere-entrained organic compounds that are dissolved in precipitation
Nitric acid	Microbial nitrification of ammonium, and leaching loss of nitrate salts (Lebedeva et al. 1979; van Breemen et al. 1982; Berthelin et al. 1985; van Miegroet and Cole 1985)	Acid rain mainly from pollutant (e.g., internal combustion engines) and natural (e.g., lightning) sources
Sulfuric acid	Microbial oxidation of sulfides in primary minerals and of reduced organic sulfur in soil organic matter generates significant amounts of sulfuric acid (van Breemen et al. 1982; Fanning and Fanning 1989). Sulfate is often soluble, especially in surficial soils, and may leach from the soil as sulfate salts	Acid rain from pollutant (e.g., coal burning) or natural (e.g., volcanic) sources (Driscoll and Likens 1982; Lindberg et al. 1986)
Carbonic acid	Respiration from microbes and plant roots. Since soil CO_2 can have 100-fold greater concentrations than that aboveground, the result is deprotonation of carbonic acid and HCO_3 leaching of bicarbonate salts that remove cations from soil (Johnson et al. 1977; Richter 1986; Amundson and Davidson 1990; Richter and Markewitz 1995b)	Precipitation's equilibration with aboveground atmospheric CO_2. A minor addition of acidity since partial pressure of aboveground CO_2 is 0.035 atm
Plant-nutrient accumulation	Uptake of nutrient ions by roots is electro-neutral and the uptake of more nutrients absorbed as cationic nutrients than as anionic nutrients results in a net release of protons to the soil system. As a result, the rhizosphere may have a relatively low pH. This acid source is particularly prominent in ecosystems with rapidly aggrading biomass (Pierre et al. 1970; Sollins et al. 1980)	Not applicable

include organic acids, microbial sulfur and nitrogen oxidation, CO_2 from soil respiration, and nutrient uptake by roots.

ORGANIC ACIDS

Mainly in surface soil horizons, organic acids have a major role in soil acidification, mineral weathering, and soil formation (Boyle and Voigt 1973; Duchaufour 1982; Buol et al. 1989; Brimhall et al. 1991; Qualls and Haines 1991). Simple organic acids, such as oxalic and citric acids, and other more complex organic acids with relatively large molecular weight, originate from metabolic products of plants, animals, and microbes (Lapeyrie et al. 1987; Herbert and Bertsch 1995). Organic acids are usually relatively low in concentration because these acids are a source of energy for microbes and can potentially be adsorbed strongly by clays. Organic acids are typically highest in concentration in O horizons and decrease sharply with depth into the mineral soil (Fox and Comerford 1990; Herbert and Bertsch 1995; Richter and Markewitz 1995b). Weathering from organic acids is most intense in surficial soil horizons where the greatest amounts of organic matter from plants are added to soil.

Organic acids are weak acids, that is, their protons dissociate incompletely, a property that depends on pH. The pH dependence of weak acids is characterized by their pK, which ranges widely for organic acids present in soil (Table 10.2). Some soil organic acids, such as oxalic, citric, malic, and formic acids, have relatively low pKs (< 4.0; Table 10.2). Acids with low pKs are readily able to contribute protons to the soil system under a wide range of pH conditions.

Many organic acids are effective ligands that enhance weathering of primary minerals by complexation of metal cations such as Al or Fe, major constituents of many primary minerals. Organo–metal complexes that are relatively soluble can be translocated downward in the soil profile.

In Calhoun soils that currently support a pine forest, dissolved organic carbon (DOC) can be taken as an index of soluble organic acids (Herbert and Bertsch 1995; Richter et al. 1995b; Markewitz et al. 1998). Concentrations of soluble DOC increase greatly as rainfall penetrates the forest canopy and infiltrates through the O horizon and upper mineral soil layers (Figures 10.1, 10.2). Concentrations of DOC are highest in water draining from the O horizon (mean over two years of 31.4 mg L^{-1}, coefficient of variation (CV%) = 28.1 among eight samplers). This DOC remains relatively high in concentration even 0.15 m into the mineral

Table 10.2. *Dissociation constants of acids that are found in soil solutions. The a1, a2, and a3 subscripts of pK refer to the sequential deprotonation of each acid (e.g., the deprotonation of carbonic acid to bicarbonate, $H_2CO_3 \rightarrow HCO_3^- + H^+$, has a pK_{a1} of 6.36). An acid's pK is the pH at which 50% of the acid's protons are dissociated*

Acid	Formula	pK_{a1}	pK_{a2}	pK_{a3}
Inorganic acids				
Carbonic	H_2CO_3	6.36	10.33	
Sulfuric	H_2SO_4	-3.00	1.99	
Nitric	HNO_3	-1.44		
Organic acids				
Acetic	$C_2H_4O_2$	4.76	$-$	
Formic	CH_2O_2	3.73	$-$	
Malic	$C_4H_6O_5$	3.40	5.11	$-$
Citric	$C_6H_7O_8$	3.13	4.76	6.39
Oxalic	$C_2H_2O_4$	1.27	4.28	$-$

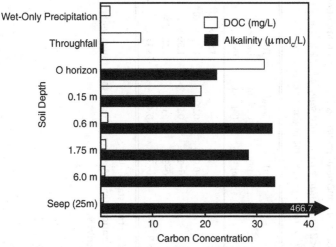

Figure 10.1. Dissolved organic carbon (DOC) and alkalinity in soil water under the loblolly pine stands of the Calhoun Forest Experiment (site P-1). The DOC is taken as an index for soluble organic acids (Herbert and Bertsch 1995). Deeper than 0.15 m depth, alkalinity is almost entirely HCO_3^-. The water samples were collected continuously from throughout the upper 6 m of soil between 1992 and 1994. Grab samples from Sparks Creek and Sparks Seep (deep drainage waters) were taken regularly (every three weeks) throughout the mid-1990s ($n = 50$).

Table 10.3. *Chemistry and microbial properties of bulk soil (conventional 6 cm-diameter core samples) in four soil horizons, and in rhizosphere soils (<2 mm distance from roots) sampled at 2 to 3 m depth in the pine-forest soil of the Calhoun Experimental Forest. Soil microbial data courtesy of Dr. Elaine Ingham, Oregon State University, Corvallis (site P-1)*

Soil material	Soil depth (m)	Total carbon (%)	Total bacteria (No. g^{-1})	FDA-active bacteria[a] (No. g^{-1})	Total fungi (m g^{-1})	FDA-active fungi (m g^{-1})
Oe horizon	–	–	1.97×10^8	32.9×10^6	59160	906
A horizon	0–0.075	0.70	1.44×10^8	23.8×10^6	18140	653
BE horizon	0.6–1.0	0.24	1.59×10^8	1.47×10^6	294	5.5
B horizon	2.0–3.0	0.073	1.23×10^8	0^b	0^b	0^b
Rhizosphere soil in B	2.0–3.0	0.42	3.17×10^8	3.54×10^6	1467	65.8

[a]FDA, fluorescein diacetate (stain).

[b]Detectable concentrations for FDA-active bacteria, total fungi, and FDA-active fungi are $< 4 \times 10^3$ units g^{-1}, < 0.3 m g^{-1}, and < 0.3 m g^{-1}, respectively.

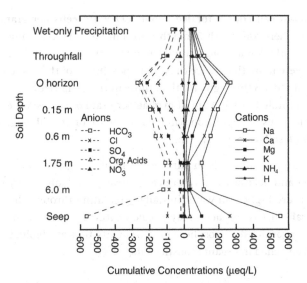

Figure 10.2. Major anion and cation concentrations of soil water under loblolly pine stands of the Calhoun Forest Experiment (site P-1). The concentrations of nitrate, sulfate, bicarbonate, and organic acids are generally indicative of the significance of each acid system.

soil (mean of 19.2 mg L^{-1}, CV% 34.4). Between 0.15 and 0.6 m, however, organic acids practically disappear from soil water. From 0.6 to approximately 25 m depth, DOC averages < 1.4 mg L^{-1} (Figure 10.1).

Despite sharply decreasing concentrations with increasing soil depth, organic acids may actually affect weathering reactions and acidification deep in the soil profile. At depth, organic acids are most concentrated in rhizospheres, i.e., soil volumes immediately surrounding roots (Berthelin and Leyval 1982; April and Keller 1990; Leyval and Berthelin 1991), and in deep fractures and solution channels (Herbert and Bertsch 1995).

Rhizospheres are important and underexplored biogeochemical environments (April and Keller 1990; Gobran and Clegg 1996), as roots and the hyphae of mycorrhizal fungi can penetrate even little weathered rock minerals (van Breemen and Buurman 1998). At the Calhoun forest, the chemistry and biology of the rhizosphere at 2 to 3 m soil depth emphasize the important role, even if localized, played by organic acids in rhizospheres deep in the subsoil (Table 10.3). For example:

- Rhizosphere soil at 2 to 3 m depth has about 6-fold the total soil organic carbon (SOC) as bulk soil at the same depth.
- Bacteria are highly concentrated in these deep rhizospheres.

Bacterial cell counts totaled about 3.17×10^8 cells per gram of rhizosphere soil, nearly 3-fold the cell count of bulk soil samples at this depth. Populations of total bacteria in rhizosphere soil at 2 to 3 m were more than twice those in bulk samples of the A horizon, and about 1.6 times greater than those in the Oe horizon.

- Fungi totaled nearly 15 m of hyphae per gram of rhizosphere soil at 2 to 3 m soil depth, compared with <0.5 m of hyphal length in all bulk soil deeper than 1 m.

In sum, organic acids in surface layers of soil greatly facilitate weathering and acidification, where concentrations of soluble organic compounds are highest in water initially percolating through the surface mineral soil. Organic acids may also be weathering agents at depth, especially in localized rhizosphere environments that are biologically active and enriched in organic constituents.

NITRIC AND SULFURIC ACIDS: PRODUCTS OF MICROBIAL OXIDATION

Nitric acid is a strong mineral acid that can weather minerals in the soil environment (Berthelin *et al.* 1985). It is produced in soils by specialized microbial populations that oxidize ammonium or organic-nitrogen substrates to nitrate in a process known as nitrification.

Nitrification can potentially generate large amounts of soil acidity. Overall, mineralization of organic nitrogen to ammonium followed by nitrification generates one mole of protons per mole of nitrogen mineralized:

$$\text{Organic N} \rightarrow NH_3 + H^+ \rightarrow NH_4^+$$
$$NH_4^+ + \tfrac{3}{2}O_2 \rightarrow NO_2^- + 2H^+ + H_2O$$
$$NO_2^- + \tfrac{1}{2}O_2 \rightarrow NO_3^-$$

In most soils, acidity from net nitrification is generated within surficial O and A horizons where organic nitrogen is most concentrated, and nitrogen mineralization is most active (Lebedeva *et al.* 1979; Berthelin *et al.* 1985; van Miegroet and Cole 1985). In a forest dominated by red alder (*Alnus rubra*) in the Cascade Mountains of Washington, USA, vigorous N_2 fixation by the symbiont *Frankia* actinomycete has accumulated large amounts of organic nitrogen, and in excess of 1 $kmol_c$ ha^{-1} of acidity is added to the soil each year from net nitrification and leaching of nitrate salt solutions. The nitric acid reacts with weatherable minerals mainly in the upper soil horizons of the Inceptisols.

Although nitrification is a potentially important process in many ecosystems (Stark and Hart 1998), as a source of acidity, net nitrification appears to be significant mainly in nitrogen-rich ecosystems. Net nitrification is not prominent in the modern forests at the Calhoun ecosystem and is not likely to be so across forests of the southeastern region as a whole (Richter and Markewitz 1995a). Most of these forests have not been fertilized with nitrogen for decades, and not only do soil solutions of the southern pine forests have low concentrations of NO_3^- (Figure 10.2), but aerobic incubations of soil exhibit low rates of nitrate production as well.

Sulfuric acid also plays a relatively small role in weathering soils at the Calhoun forest. Sulfuric acid is a strong mineral acid (pK ~ -3.0), and is produced by microbial oxidation of reduced sulfur at relatively low redox potentials. Sources for oxidized sulfate produced in soil include reduced sulfur in amino acids, elemental sulfur, and sulfide. Because of the relatively low redox potential for microbial sulfur oxidation to sulfuric acid, sulfur oxidization may be found in microsites throughout many soil profiles, from O horizons to the bottom of the C horizon, and even within the underlying bedrock itself. A significant source of acidity in many soil profiles may be derived from the microbial oxidation of the sulfide that is present within primary minerals.

A reaction that describes sulfuric acid production from oxidative weathering of pyrite, FeS_2, an oxidation that generates four moles of protons per mole of pyrite, is as follows:

$$FeS_2 + 3\tfrac{3}{4}O_2 + 3\tfrac{1}{2}H_2O \rightarrow Fe(OH)_3 + 2H_2SO_4$$

Biogenic sulfur oxidation of C horizons is well documented in Ultisols derived from sediments on the Coastal Plain of Maryland, USA by Fanning and Fanning (1989), who suggest that sulfide oxidation is underestimated as a source of mineral weathering in deep layers of many soils.

The significance of these sulfur reactions depends on the concentration of reduced S in the original geologic material and the redox conditions of the soil environment. Since total sulfur exists only in small to trace amounts in the unweathered granitic gneiss that provides the mineral material for the Calhoun soil profile, we suggest that sulfur oxidation contributes only secondarily to the acidification that has occurred to form the Calhoun's advanced weathering-stage soil. What acidification results from sulfur oxidation is most likely concentrated deep in the C horizon.

Not all sulfuric acid is derived from within-ecosystem sources, especially over time scales of multi-millennia. Sulfuric acid that is

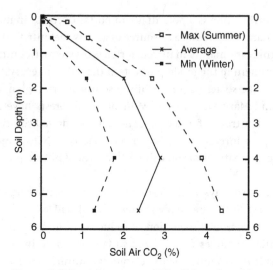

Figure 10.3. Partial pressure of CO_2 (% by volume) increases substantially with depth as a function of both soil respiration and reduced diffusivity. The resultant pattern of CO_2 pressures is responsible for intensive weathering of soil minerals at depth, potentially even below the depth of active rooting. Site P-1.

deposited from the atmosphere affects mineral weathering. Downwind of emissions of SO_2 from industrial, volcanic, or other sources, atmospheric deposition of pollutant S can contribute substantial amounts of acidity to soil and accelerate current mineral weathering. Effects of atmospheric deposition of pollutant acidity are considered in more detail in Chapter 17, which examines changes in Calhoun soil acidity over time scales of decades.

CARBONIC ACID

Respiration is a central process of ecosystems, and organic matter decomposition and plant-root respiration elevate soil CO_2 greatly. Soil's partial pressure of CO_2 in the soil atmosphere controls the significant weathering processes affected by carbonic acid (Reuss and Johnson 1986; Amundson and Davidson 1990; Richter and Markewitz 1995b). Carbonic acid weathering is a three-phase system that includes CO_2 in the gas phase, carbonic acid and associated ions in the liquid phase, and the solid phase of minerals and cation exchange surfaces. The system weathers minerals throughout soil profiles, but since partial pressures of CO_2 are most elevated at depth (Figure 10.3), B and C horizons are subject to the most intense effects of this weathering system.

Table 10.4. *Solution pH of low ionic strength solutions in equilibrium with CO_2 at different partial pressures. The soil atmosphere > 1 m depth at the Calhoun ecosystem may exceed 5% CO_2, and atmospheres of some soils approach and exceed 10%*

CO_2 (%)	pH	HCO_3^- (mmol L^{-1})	$H_2CO_3^*$ (mmol L^{-1})
0.036	5.65	0.0029	0.017
1.0	4.9	0.0145	0.479
5.0	4.6	0.036	2.39
10	4.4	0.046	4.79
100	3.9	0.145	47.9

Reactions are detailed in Stumm and Morgan (1981) and are summarized in Table 10.4 and Appendix I. This information helps illustrate one of the most interesting aspects of Calhoun soil acidification. Despite extreme soil acidity of nearly the entire soil profile (Table 9.1), the weak carbonic acid is able to continue to supply protons to the already acidic soil (Richter and Markewitz 1995b).

On first examination, since carbonic acid is a very weak acid (pK_{a1} of 6.36), the great acidity of the Calhoun soil would appear to diminish the potential for carbonic acid to contribute protons to the soil, since soil has a pH well below the pK of carbonic acid. On the other hand, CO_2 ranges between 1 and 5% in soil air below 1.75 m depth at the Calhoun forest (Figure 10.3), and these elevated partial pressures of CO_2 ensure high concentrations of $H_2CO_3^*$ (dissolved and hydrated CO_2) in solution and that protons of a small fraction of $H_2CO_3^*$ will dissociate, despite low solution pH. Dilute HCO_3^- is measurable (Figure 10.2), and Table 10.4 summarizes equilibrium calculations that demonstrate that at 1 or 5% CO_2, *in situ* pH of soil waters can be depressed to 4.9 or 4.6, respectively. Under these conditions of greatly elevated CO_2 and dilute electrolyte concentrations, HCO_3^- is calculated to be about 15 and 35 μmol L^{-1} in soil water, very close to what was measured by titration (Figures 10.1, 10.2) in two years of soil-water collections (Markewitz *et al.* 1998).

Annual hydrologic flux of HCO_3^- at 6 m depth of Calhoun soils is estimated to range from 0.11 to 0.22 kmol$_c$ ha^{-1}, based on the linkage of two years of chemical measurements of drainage-water alkalinity and a hydrologic simulation by the PROSPER model (Gnau 1992; Richter *et al.* 1994; Markewitz *et al.* 1998). This flux is the rate of carbonic-acid-driven acidification of the soil system, from 0 to 6 m (Binkley and Richter 1987; Johnson et al. 1991; Markewitz et al. 1998).

Carbonic acid dynamics of solutions that drain through material at >6 m depth are as interesting as those at <6 m. Bicarbonate concentrations of soil water at 6 m depth averaged 33.4 μmol L^{-1} ($\pm$ 2.89 standard error), but by the time the water enters Sparks Creek, about 25 m vertical distance below the Calhoun forest soil surface, it has gained nearly 10-fold more HCO_3^- (Figure 10.1). When viewed from the perspective of the full soil profile and weathering zone, soil at <6 m depth continues to be acidified slightly by the carbonic acid system (at 0.11 to 0.22 kmol ha^{-1} $year^{-1}$), but the effects of the carbonic acid system's acidification and weathering have far more impact (perhaps up to 10-fold greater) on soil-water chemistry at >6 m (Figures 10.1, 10.2). Effects of soil respiration are seen to penetrate very deeply into the weathering profile.

Although the relatively high pK of carbonic acid limits the acid's ability to lower solution pH to less than 4.5 and mobilize Al from the soil profile (Reuss and Johnson 1986), elevated subsoil CO_2 ensures that carbonic acid affects desilication and can create an Al-saturated soil.

NUTRIENT UPTAKE AND ACCUMULATION BY VEGETATION

A fifth source of acidity is derived from the growth of vegetation, as a consequence of root uptake of nutrient cations and anions. The physiological process of root uptake of nutrients is charge balanced, and uptake of cations and anions affects a release of H^+ and OH^-, respectively, into the surrounding soil environment. The accumulation of nutrients in plant biomass acidifies mineral soil if the biomass contains more nutrients taken up as cations than as anions. The mineralization of organic-bound nutrients reverses the process. A variety of recent scientific literature describes details of the acid balance of terrestrial ecosystems, including old-growth forests, aggrading secondary forests, and cultivated field crops (Pierre et al. 1970; Sollins et al. 1980; Ulrich et al. 1980; Driscoll and Likens 1982; van Breemen et al. 1982; Binkley and Richter 1987; Johnson and Lindberg 1992; Markewitz et al. 1998).

Not only does root uptake of nutrients affect the soil's acid balance, but within the soil rhizosphere, ionic uptake phenomena can stimulate mineral weathering. Extremely low pH has been measured in the rhizosphere, and April and Keller (1990) indicate that free protons in rhizospheres may greatly stimulate mineral weathering.

Over time, it is not simply root uptake but also the cation and anion accretion in plant biomass that affects soil acidification (Richter 1986). For example, as glaciers melted during the Holocene, plant biomass and soil organic matter increased greatly (Harden *et al.* 1992). This primary succession also acidified soils as nutrient cations were probably accumulated by vegetation in much greater quantities than anions. Effects of nutrient-ion uptake on soil acidity are most pronounced in ecosystems with aggrading or degrading biomass, a process most active over time scales of decades and centuries. We return to these ion-uptake issues in a subsequent chapter that describes the liming and re-acidification of the Calhoun soil (Chapter 17).

AN OVERVIEW OF THE ACID-WEATHERING ATTACK

Protons derived from mineral or organic acids or generated within the rhizosphere during nutrient-cation uptake participate in one of two soil reactions. The first reaction is proton exchange with the solid phase, acidifying soil as protons displace nutrient cations such as Ca^{2+} that are electrostatically adsorbed to the soil. Cation exchange reactions are summarized here by using protons from the carbonic acid system to displace Ca^{2+} to soil solution which drains from the soil system:

$$2H_2CO_3^* + \text{soil Ca} \rightarrow \text{soil } 2H^+ + Ca^{2+} + 2HCO_3^-$$

where the reaction products are acid-saturated soil that may spontaneously decompose to an Al-saturated exchange surface in mineral soils (Thomas and Hargrove 1984). Under humid conditions, the calcium bicarbonate salt will leach and potentially be translocated within the soil profile or lost to ground and surface waters. Similarly, H^+ from $H_2CO_3^*$ may displace exchangeable Mg^{2+}, K^+, and Na^+ from soil which may be leached as bicarbonate salts in drainage water.

In the second reaction, protons derived from biogenic acids react with soil minerals in weathering reactions which can be congruent or incongruent. In congruent weathering, acids react with a mineral surface and all reaction products enter solution and may be leached from soil to groundwater or nearby streams. Two examples of congruent weathering are the decomposition of serpentine and calcite, respectively:

$$Mg_3Si_2O_5(OH)_4 + 6H_2CO_3^* \rightarrow 3Mg^{2+} + 6HCO_3^- + 2SiO_2 + 5H_2O$$
$$CaCO_3 + H_2CO_3^* \rightarrow Ca^{2+} + 2HCO_3^-$$

where all reaction products are soluble and translocated by leaching through the soil.

In incongruent weathering, acids react with the mineral surface but reaction products are only partly leached from soil to drainage water as products are recombined in secondary minerals within the soil profile. An example is the dissolution of albite, $NaAlSi_3O_8$, a prominent feldspar and a common primary mineral in granites:

$$2H_2CO_3{}^* + 2NaAlSi_3O_8 + H_2O \rightarrow$$
$$Al_2Si_2O_5(OH)_4 + 2Na^+ + 2HCO_3{}^- + 4SiO_2$$

where reaction products are the insoluble kaolinite, $Al_2Si_2O_5(OH)_4$, plus soluble products of sodium bicarbonate salt and silica, both of which can leach from the soil. In this incongruent reaction, all of the Al and part of the Si in albite re-precipitate to form the secondary mineral kaolinite.

A MILLENNIAL PERSPECTIVE OF CALHOUN ACID SOURCES

In the Calhoun soil and ecosystem, a general hypothesis is presented of the relative significance of acid sources that have affected its extreme soil acidity (Table 9.1). To describe this hypothesis of acid generation, we subdivide the soil profile into upper and lower soil systems (Brimhall *et al.* 1991; Richter *et al.* 1995b). The upper system includes the O, A, and B horizons (the horizons of the traditional *solum*). The lower soil system includes the C horizon or saprolite.

Over millennial time scales, organic and nitric acids have contributed significant amounts of acidity to the Calhoun surface soils. Both acids are closely associated with organic matter decomposition, which is usually most concentrated in the most surficial soil layers. Organic acids have had important effects in the lower soil system, specifically in deep rhizospheres and in fractures and solution channels as well. Acidity due to root uptake of nutrient ions probably follows a distribution similar to that of the organic acids, affecting the upper soil system, and having an important if secondary role in rhizospheres of the lower system as well. This latter acidity is contributed to the profile depending on root distribution and spatial patterns of nutrient uptake. At Calhoun, roots have been observed at 4 m depth and we presume that they penetrate more deeply. In addition, since the spatial distribution of roots is dynamic through time, rhizospheres may over time potentially affect a large fraction of the soil's volume.

By contrast, sulfuric acid has potentially acidified the soil profile in both upper and lower systems, as well as in the underlying bedrock itself. Sulfide-oxidizing bacteria are able to acidify soil throughout the profile, whereas sulfuric acid from atmospheric deposition (whether ancient or modern) mainly acidifies surface soils. The concentration of sulfide in unweathered granitic gneiss at the Calhoun forest is low, so that soil sulfide oxidation is probably a secondary contributor of soil acidity.

Lastly, carbonic acid is positively correlated with soil depth since it is controlled by CO_2 concentration which is strikingly depth dependent (Figure 10.3). Carbonic acid is influential in both upper and lower soil systems although the concentration gradient of CO_2 is frequently very steep within the upper system and the concentration of CO_2 is often low in surface horizons due to high diffusivity of A horizons. Pronounced seasonal fluctuations of soil CO_2 are observed in the soil profile, with concentrations varying in the upper C horizons between 1% in the dormant season and $>5\%$ in the growing season (Figure 10.3).

In concert, the ecosystem's biogenic acids attack soil minerals throughout the soil profile. Chemical elements are released and taken up by plants and microbes to meet nutritional requirements, adsorbed to electrostatically charged surfaces, complexed by ligands, recombined into secondary clay minerals, or leached to groundwaters, rivers, wetlands, lakes, and eventually to the ocean. At Calhoun, the upper soil system is subject to a battery of acids from the oxidation of carbon, nitrogen, and sulfur, root uptake, and organic acids. In the lower system, acids are mainly derived from carbonic acid and secondarily from sulfuric and organic acids, with relatively small contributions from plant-nutrient uptake.

On stable *terre firme* landforms, the ultimate products of acidification and weathering are advanced weathering-stage soils, such as Ultisols and Oxisols (Table 1.1). Only a few chemical elements, such as Zr and Ti, are insoluble enough to resist transportation from weathering environments (Table 6.1). The depauperate Calhoun soil described in Table 9.1 illustrates the result of this multi-millennial acidification process. Since these soils are composed of the most insoluble chemical compounds, their ability to sustain a balanced supply of nutrients in bioavailable form will always be to some extent in question.

Part III

Soil change over time scales of centuries: conversion of primary forests to agricultural fields

"Agriculture <u>was</u> the South, economically and in environmental impact up until the Civil War, and it remained a major influence thereafter."

A.E. Cowdrey (1996)

11

Agricultural beginnings: Native American cultivation

Over the last three centuries, southeastern North America has been one of the world's major agricultural regions. Much of this agricultural economy has been based on the advanced weathering-stage Ultisols that extensively cover the region, soils such as those found at the Calhoun forest.

The origins of southeastern agriculture are, however, many centuries older than the 1700s, and we can learn much about land-use effects on soil change from Native Americans' cultivation of maize (*Zea mays*) throughout the region. Maize cultivation in eastern North America is recognized as one of the world's four independent centers of agriculture, the others being the Near East, northern China, and Mesoamerica (Smith 1989).

This chapter briefly examines maize cultivation of southeastern Native Americans, because it can suggest much about ecological processes that sustain and deplete soil's nutrient supply. This agriculture was based almost exclusively on the region's highly fertile Inceptisol soils of alluvial bottomlands, rather than on the far more extensive acidic Ultisols of the uplands. We conclude the chapter by looking at a hypothetical nitrogen budget of a maize management system which suggests how these systems were able to maintain their productivity over many centuries.

SOILS AND THE MAIZE CULTURE

One of the most significant achievements of Native Americans was the development of a system of maize (*Zea mays*) agriculture, which in southeastern North America was continued on a large scale for centuries, from as early as 800 to 1700 AD. Native Americans were the first to manage what are now many of the world's most important crops: not only corn, but also beans (*Phaseolus* spp.), potatoes (*Solanum tuberosum*), tomatoes (*Lycopersicon esculentum*), cotton (*Gossypium* spp.), tobacco (*Nicotiana tabacum*), peppers (*Piper* spp.), squash and gourds (*Cucurbita pepo* and *Largenaria silceraria*), and sunflower (*Helianthus annuus*).

Indigenous agricultural ecosystems in southeastern North America were initially developed with local plant domesticates that did not include corn. But without question, it was the Native Americans' widespread cultivation of corn (an exotic plant from Mesoamerica) that had the most impact on humanity, both in the ancient Native American world and in the modern world.

Native Americans greatly improved the genetic potential of corn over time scales of centuries (Hudson 1976; Johannessen and Hastorf 1994). Early varieties of corn arrived from tropical America and were first cultivated in eastern North America about 2000 years ago. Corn was grown as a minor crop for many centuries prior to its widespread adoption as a primary food source throughout much of eastern North America between about 800 and 1100 AD. The Native Americans grew all the major types of corn known today: Flint, Flour, Pop, Dent, and Sweet corn (Wallace and Brown 1988). The shift to maize as a dietary staple can be documented by increases in $^{13}C/^{12}C$ ratios of human bone collagen that closely match the unusual $^{13}C/^{12}C$ ratio of corn (Ambrose 1987; Schoeninger and Schurr 1994).

Productivity of corn in the southeast increased greatly after about 800 AD (Scarry 1993). By 1100 AD, the cultural development known as the Mississippian tradition had become dependent on corn cultivation. Hudson (1976) suggests that at its apex the Mississippian tradition represented the highest cultural achievement of indigenous peoples in all of North America.

Corn grew well in the moist alluvial soils and warm temperate climate of southeastern North America. Mississippian populations burgeoned in large part due to increased maize production. Some Mississippian towns had tens of thousands of inhabitants, and included many networks of farmsteads scattered along nearby alluvial floodplains and terraces (Smith 1985). The towns were well organized (Figure 11.1), and protected by wooden stockades and moats that surrounded homes, gardens, ceremonial buildings, and mortuaries. Large earthen mounds were constructed of soil piled by hand, sometimes 30 m high and hundreds of meters in width and length (Kennedy 1996).

The Native American corn fields and granaries impressed early European visitors to the region, including, for example, the Spaniards of

61. "THE TOVVNE OF SECOTA." Engraving number 20 from De Bry, 1590, 9 x 12¼ inches. After White. North Carolina, 1585, *NYPL*.

Figure 11.1. De Bry's 1590 engraving "The Tovvne of Secota" in south-eastern North America, illustrating an agronomic community (Fundaburk 1958).

Hernando de Soto's expedition in the 1540s (Bourne 1904) and the young John Lawson from England in 1700 (Lefler 1967). The 600-man *entrada* of de Soto depended on the corn they obtained from Native American stores by barter, theft, and ransom (Bourne 1904). Even in spring, the Spanish located large quantities of corn in storage granaries, illustrating the potential productivity of corn cultivation.

Table 11.1. Chewacla soil-chemical properties. These bottomland terrace soils were sampled in old mixed bottomland forests along the Tyger River in South Carolina on sites cultivated or potentially cultivated by Native Americans of the Mississippian tradition. These were also managed for crops for most of the period of the early 1800s to about 1930. Table 9.1 describes analytical methods

Horizon or depth (m)	C (%)	N (%)	C/N	Mehlich III PO_4-P (μg g^{-1})	Exchangeable (mmol$_c$ kg^{-1})					BS (%)
					Ca	Mg	K	Acidity	ECEC	
Chewacla at the Old Ray Place										
O horizon	18.05	0.771	23.4	–	202.6	60.4	16.7	3.04	28.37	98.9
0–0.3	2.56	0.213	12.0	7.26	60.6	20.4	2.0	1.80	8.62	97.9
0.3–0.55	1.00	0.085	11.8	2.24	30.5	16.7	1.0	4.13	5.40	92.3
0.55–0.875	0.59	0.045	13.3	2.55	16.5	13.9	0.4	3.48	3.63	90.4
0.875–1.10	0.33	0.027	12.0	2.50	8.1	14.3	0.3	3.43	2.92	88.2
Chewacla at Rose Hill										
O horizon	12.05	0.560	21.5	–	99.2	38.8	9.8	5.29	15.43	96.6
0–0.3	2.71	0.220	12.3	11.60	23.7	15.5	1.6	16.61	5.85	71.6
0.3–0.55	1.17	0.107	10.9	4.53	32.1	29.6	1.2	1.98	6.62	97.0
0.55–0.875	0.72	0.060	12.0	5.39	31.5	27.9	1.0	0.94	6.26	98.5
0.875–1.10	0.42	0.038	11.1	0.38	41.6	36.3	0.8	6.08	8.61	92.9

John Lawson recorded observations of corn in his well documented trip by canoe and foot through the Carolina Coastal Plain and Piedmont in the winter of 1700–1701. Lawson was most impressed by productive corn fields on river flats and by corn-based meals, especially loblolly stew, to which he was treated as an honored visitor while traveling among the Congarees, Waterees, Saponas, Enos, and other Native American tribes (Lefler 1967). Other impressive descriptions of the prolific maize culture are found in Swanton (1946), Hudson (1976), and Smith (1989).

HOW AND WHY THE CORN ECOSYSTEM WORKED

Several features of Native American corn cultivation ensured that the soil-management system worked well over time scales of centuries. These included the high inherent fertility of alluvial soils (Table 11.1), interplanting with legumes, use of fire, and the relatively modest soil-physical impacts of this manner of cultivation. Periodic fallows may also have been important to the continued management of these alluvial soils, due to excess weeds, build-up of crop pests and pathogens, or reduced soil fertility.

First, it is no coincidence that corn cultivation in the region was almost entirely practiced on alluvial soils of river bottoms (Figure 9.2) rather than on the acidic, residual soils of the uplands. Many terrace soils in southeastern North America have high inherent fertility which made these excellent sites for agriculture. Compared with the upland acidic Ultisol described in Table 9.1, the alluvium cultivated by Native Americans is fertile by nearly every measure (Table 11.1). Even at 1 m depth, total carbon and nitrogen, exchangeable calcium and magnesium, and extractable phosphorus in alluvial Chewacla soils exceed concentrations in the upland Appling soils by 3-, 3-, 100-, 10-, and 25-fold, respectively (Tables 9.1, 11.1). On such soils, crop productivity is relatively easy to maintain. These river terraces are mainly composed of Inceptisol soils that often have a deep but reliable source of soil moisture and some of the highest potential productivities in all of North America.

In an evaluation of soil fertility and potential productivity along the Black Warrior River near Moundville, Alabama, Peebles (1978) estimated that indigenous agriculturists produced corn on some soils that yielded about 0.6 to 2.4 Mg ha^{-1} (10 to 40 bushels per acre), high rates of production considering that these systems had relatively low management inputs. Muller (1978) describes how at the Kincade settlement along the Tennessee River, maize growing on soils with mid-20th century technology produced 5 to 7.5 Mg ha^{-1} of corn (80 to 125 bushels per acre).

Second, nitrogen-fixing beans (*Phaseolus* spp.) were often inter-planted with maize, and may well have increased nitrogen inputs to the ecosystem. Annual harvest removals of nitrogen in corn grain may have made nitrogen inputs important to the biogeochemical sustainability of the system. Rates of nitrogen fixation under these conditions are known with little certainty, and leguminous fixation rates are also highly vari-able among legume species. Perhaps as much as 10 to 30 kg ha^{-1} of nitrogen might have been annually added to the soil system apart from that harvested in the bean crops themselves. Cumulatively, these inputs may have been significant to the continued productivity of the corn ecosystem.

Third, fire was almost certainly used to control weeds and brush and to mineralize nutrients. Although nitrogen and sulfur are lost to the atmosphere during biomass burning, fires kill weeds and recycle organic-bound ash elements such as phosphorus, calcium, potassium, and magnesium to bioavailable forms. Control of competing vegetation by fire would enhance crop uptake of nutrients and water.

Fourth, adverse physical effects of corn management on the alluv-ial soils were modest. The cultivation probably did not greatly reduce soil organic matter, given that the soil was disturbed only with sticks, animal bones, and shells (Figure 11.2). These implements broke the soil surface but hardly mixed the upper volume of soil like a mule- or tractor-drawn plow. Maintenance of soil organic matter helped ensure that soils did not lose water storage capacity and nutrient-retention ability, or experience elevated toxicity of Al and rapidly diminishing rates of nutrient mineralization. Like minimum-till systems today, culti-vation by Native Americans probably increased erosion by only a small fraction of that created by modern conventional tillage. Erodibility factors for soil under modern minimum-till systems are 20% or less of those for soil under modern conventional tillage.

THE HYPOTHETICAL MAIZE-NITROGEN CYCLE

Although the inherently high fertility and reliable water supply made these alluvial soils ideal for agricultural management, fallows may have been used as a method to cope with pathogens, competing weeds, and even diminished nutrients. Nitrogen removals due to crop harvests and fire may have significantly affected maize yields. By contrast, harvest removals of phosphorus, potassium, magnesium, and calcium probably had modest impacts on nutrient bioavailability. Harvest removals of these nutrients were relatively low compared with nutrient capital already within the systems.

Table 11.2 details a hypothetical nitrogen budget for an alluvial soil cropped with maize. The budget suggests that some maize fields required periodic fallows as a practice to rebuild mineralizable soil nitrogen.

In the maize system, nitrogen is lost via several processes: in the harvest of corn and other crop products, by fire-caused oxidation, and by soil leaching and denitrification of nitrate. Annual losses may have totaled as much as 25 to 65 kg ha^{-1} as nitrogen (Table 11.2). With no nitrogen amendments from fertilizers or imported organic matter (Hudson 1976), soil-nitrogen supply depended mainly on continued mineralization of soil organic N. Except for nitrogen contained in periodic alluvium, organic nitrogen was supplemented only by fixation inputs from legumes (i.e., from interplanted beans) and by inputs of nitrogen from atmospheric deposition. These inputs are estimated to have ranged from 5 to 35 kg ha^{-1} as nitrogen. From the budget in Table 11.2 it appears that mineralizable organic nitrogen, even on fertile alluvial sites, might have been slowly depleted.

Although Table 11.2 suggests that fallows may have been necessary to renew bioavailable nitrogen, cycles of cropping and fallow in these maize systems are known with little certainty. Fallows are used in shifting agricultural systems throughout the world in response to the wearing-off of fertilizer effects from fire, excess weed competition, pests, and pathogens. Nonetheless, the rate of natural vegetative regrowth indicates that the nitrogen economy of alluvial soils might have been readily renewed following cropping for corn. Over a 10-year fallow, for example, the ecosystem could have accumulated 20 to 200 kg ha^{-1} of new soil nitrogen (Table 11.2), not including the nitrogen recycled and accumulated in the 20 to $>$ 60 Mg ha^{-1} of new plant biomass and detritus of fallows. The system could be chopped, burned, and trees girdled to re-prepare the soils for corn and other crops.

Following a fallow period, clearing and burning of new biomass and organic detritus fertilized surface soils. Fire releases organic-bound nutrients to soluble and bioavailable forms, and after burning, rainfall leaches nutrients contained in ash into the root zone. Such post-burn chemical reactions raise soil pH and enhance the bioavailability of nutrients such as phosphorus, potassium, magnesium, and calcium. Bioavailability of nitrogen may also have been enhanced due to increased soil pH and release of ash elements. Woody brush may have been

Table 11.2. *Hypothetical nitrogen budget of Mississippian maize supported by Chewacla soil in the Carolina Piedmont during years under harvest and fallow*

Loss or input	N budget under harvest (kg ha^{-1} year^{-1})	N budget under fallow (kg ha^{-1} year^{-1})	Comments and approximations
Nitrogen losses			
Harvest	15–30	0	Removals in harvested corn grain yields (at 3 to 4% N) at 0.4 to 1 Mg ha^{-1}
Annual fire	10–30	0	Combustion of 2 to 6 Mg ha^{-1} organic matter with nitrogen of 0.25 to 0.75%
Leaching	1–5	<1	Drainage of 50 cm year^{-1} with 0.2 to 1 mg L^{-1} nitrogen: during fallows nitrogen decreases to <0.2 mg L^{-1}
Total losses	25–65	<1	
Nitrogen inputs			
Atmospheric deposition	1–2	1–2	Pre-industrial nitrogen deposition
N$_2$ fixation	5–20	<1–5	Free-living plus symbiotic leguminous fixation
Net sediment deposition	1–15	1–15	Deposition rates 0.1 to 0.5 mm year^{-1}, with total nitrogen at 500 to 3000 µg g^{-1}
Total inputs	5–35	2–20	
Net system flux	+ 10 to − 60	+ 2 to + 20	The cropped system is likely to have been depleted of nitrogen on many sites; fallows that accumulate organic matter may help site recover mineralizable N

Figure 11.2. Native Americans in Virginia planting fields for corn, from the De Bry 1591 engraving entitled *Mode of Tilling and Planting* (Fundaburk 1958).

piled around larger trees to more effectively burn debris and to facilitate its decomposition.

In sum, key processes of these maize systems in southeastern North America that sustained soil fertility were high inherent fertility of the alluvial soils, periodic sediment deposition, symbiotic N_2 fixation, mineral weathering, atmospheric deposition, and probably periodic fallows. At least three of these factors are unique to the alluvial terrace soils and thus are not available to supporting agriculture on the uplands, as we will see in the next chapter.

12

Soil biogeochemistry in cotton fields of the Old South

116   To use soil without damaging it is nowhere easy. This was certainly the case in southeastern North America, as European- and African-Americans expanded farmland from the Native American-cultivated alluvium on to the region's uplands during the 18th and 19th centuries. The effects of this expanded cultivation were so great that the region's entire soil resource has been transformed from what it was prior to clearing and being converted to agriculture. Even still, few cannot be impressed with the prodigious efforts that went into establishing one of the world's leading agricultural economies. The USA South from Virginia to east Texas led the new nation in agricultural production starting early in the 19th century. The region retained that status through much of that century, and agricultural production in many parts of the region remains high even today.

In this and the following chapter we evaluate the biogeochemical transformation of the region's soils as they were affected by this agricultural development.

EXPANDED CLEARING AND FOREST CONVERSION

Across much of southeastern North America, many of the first fields cultivated by the new pioneers were the same alluvial lands that had been cropped for maize by Native Americans. Journals, letters, and other records describe how early settlers were anxious to cultivate arable, alluvial Inceptisols already used by Native Americans.

Early pioneer agriculture was not dissimilar to indigenous agriculture. Not only were methods of clearing and cultivation akin to those of Native Americans, but early pioneers' diet, clothing, and subsistence hunting shared much with the Native American world, a world that has now been relegated to novels and museums.

Figure 12.1. Basil Hall's (1829) etching entitled *Newly Cleared Land in America*. The scene is of western New York state, but it can represent all recently cleared forests in eastern North America in the early 19th century. Hall's caption reads: "The newly cleared lands in America have almost invariably, a bleak, hopeless aspect. The trees are cut over at a height of three or four feet from the ground, and the stumps are left for many years till the roots rot; – the edge of the forest, opened for the first time to the light of the sun, looks cold and raw; – the ground, rugged and ill dressed, has a most unsatisfactory appearance, as if nothing could ever be made to spring from it ... As land is of little value, no care is taken to limit the width of roads, which are often 20 or 30 yards broad, along which carriages may find their way as they best can. The whole scene has no parallel in old countries."

Native American corn fields were soon growing new settlers' crops. Well drained alluvium was extensively cleared. But very rapidly, the upland forest was cut to make way for new cropland. The great agrarian push into the vast upland forest had begun.

Tocqueville observed forest clearing in the USA in the early 19th century (Mayer 1960) and described the difficulty of clearing and domesticating the frontier's forest. Basil Hall's (1829) etchings of similar scenes (Figure 12.1) give additional detail to Tocqueville's words:

> [T]he whole country has the look of a wood in which clearings have been made ... Every sign of a new country ... Man still making clearly ineffective efforts to master the forest. Tilled fields covered with the shoots of trees; trunks in the middle of corn. Nature vigorous and savage.

Clearing the forest was incredibly laborious. In addition to tree-by-tree felling, girdling and fire were the main methods by which forests

Table 12.1. *Estimated nutrient content of biomass of mature deciduous forests in the temperate zone (Marion 1979). Most of the phosphorus, potassium, calcium, magnesium, and micronutrients remained on site following clearing, whereas at least some of the nitrogen and sulfur was lost by oxidation during burning. Median biomass of forests is estimated to be about 340 Mg ha^{-1} with a range of 147 to 504 Mg ha^{-1}. Effective nutrient concentrations in 0.3 m soil are estimated to illustrate equivalent soil concentrations compared with those in unfertilized upland hardwood forests (from Table 9.1). Such comparisons illustrate the potential for soils to be enriched following clearing and burning of the original forest*

Nutrient	Median in biomass of mature forests (kg ha^{-1})	Range in biomass of mature forests (kg ha^{-1})	Effective concentration in soil[a] from biomass nutrients	Observed concentration in hardwood forest soil[b]
N	1085	406–1608	135[c]	400[c]
P	73	36–99	18.7[c]	2.4[c]
K	463	286–531	3.0[d]	0.6[d]
Ca	1142	644–1334	14.6[d]	0.2[d]
Mg	115	93–123	2.5[d]	0.3[d]
S	76	70–82	–	–
Fe	27	–	0.4[d]	–
Mn	125	–	1.2[d]	–

[a]Concentration assumes 0.3 m surface mineral-soil depth (for comparison with Table 9.1) and a bulk density of 1.3 g cm^{-3}. For nitrogen, burning is taken to transfer 50% to the atmosphere.
[b]Estimated from unfertilized hardwood forest soil in Table 9.1.
[c]In μg g^{-1}.
[d]In mmol$_c$ kg^{-1}.

were leveled. Downed wood was used for fences, building construction, fuelwood, and fertilizer (created by burning slash). Most forest biomass and debris in the southern forest probably remained on site although it was converted to partially burned organic matter and ash (Figure 12.1).

Although at least some of the nitrogen and sulfur would have been lost from the ecosystem following clearing and fire, nearly all of the remaining nutrients would have been added to the mineral soil and be readily available for uptake by crops (Table 12.1). Such inputs were significant to the productivity of the first generations of crops, especially considering the lack of available fertilizers. A comparison of soil concentrations in Table 12.1 with those in native hardwood forest soil in Table

9.1 illustrates the substantial input of nutrient elements potentially available in the biomass of the original forest.

During the first half of the 19th century, a shifting-field cultivation was at least occasionally practiced across the region. As fertilizer effects of clearing wore off, crop productivity diminished. Fields were then abandoned and "fresh soil" was brought into cultivation. Shifting fields were reminiscent of Native American approaches to agriculture, although the significant difference was that the majority of the new arable uplands were Ultisol soils, far more depauperate soils than the Inceptisols of the alluvial bottoms previously cultivated by Native Americans.

This method of cultivation developed in response to the times. Land was relatively cheap, the rural economy was based on slavery, population pressures were not great, and fertilizers were not extensively available prior to the Civil War (Sheridan 1979). The labor of slaves provided the means to shift spent fields to fresh soil, and at least occasionally to bring previously cultivated fields back into cultivation. Due to free-ranging livestock, manure was neither collected nor concentrated in barns.

Across the region, when crop yields decreased, farmers tended to move on to "fresh soil" as the main management approach to problems with pathogens, pests, weeds, nutrient losses from harvests, and accelerated decomposition, erosion, and leaching. Durations of crop and fallow cycles ranged widely, depending on local conditions and market forces. From 10 to 25 years was probably a minimum time for weeds and trees to accumulate significant amounts of biomass and nutrients that could be released in the upper mineral soil by clearing, chopping, girdling, and burning.

PRE-WAR COTTON: 1790S–1860

After the United States became an independent nation, the conversion of southeastern forests to row-crop agriculture was rapid. Forest conversion was driven by relatively high prices for agricultural products and plentiful land. The focus of many southern farmers was the export market, initially with rice, indigo, silk, and tobacco. Following Whitney's invention of the cotton gin in the 1790s, lint could be efficiently separated from seed, and thereafter cotton dominated the region's agriculture.

Throughout the 1700s, cotton had been grown in small farm fields and gardens, mainly for on-farm use. Before the cotton gin, management of cotton had not changed much since its independent origins in

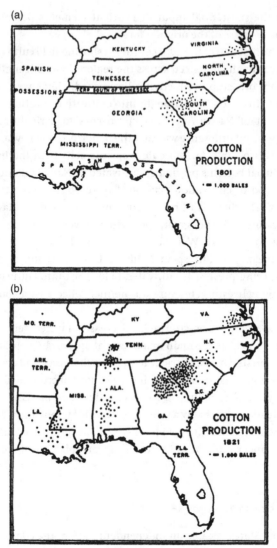

Figure 12.2. Gray's (1933) maps of the expansion of cotton production across the southeastern USA.

ancient America and the Near East. Whitney's invention, however, rapidly transformed cotton from a garden crop into a major international commodity. Cotton dominated much of the southern United States' ecology, culture, and economics until the early 20th century.

The expansion of cotton (Figure 12.2) illustrates the rapid geographic expansion of southern agriculture (Gray 1933). Cotton cultivation originated in South Carolina, and for many years the Piedmont of

(c)

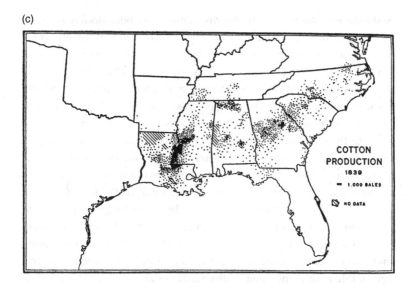

(d)

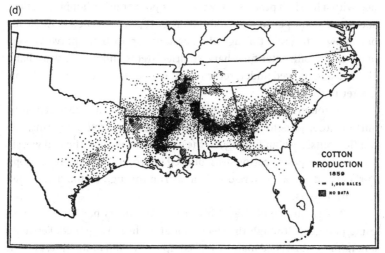

South Carolina and Georgia was the center of the expanding cotton economy (Figure 12.2). By 1801, South Carolina produced 20 million pounds of cotton; by 1811 about 40 million. The entire national production in these two years totaled 40 and 80 million pounds, respectively.

By the 1820s, cotton was still grown mainly in the Piedmont of South Carolina and Georgia (Figure 12.2). By the 1840s the crop continued to expand in South Carolina and Georgia, but it had exploded across large areas of Alabama, Louisiana, and Mississippi. By this time, cotton had become the predominant crop across the Black Prairie of

Alabama and Mississippi and on Mississippi River bottoms (Figure 12.2). By the beginning of the Civil War in 1861, cotton was extensively grown throughout the South, from the Piedmont of South Carolina and Georgia to east Texas, Arkansas, and Louisiana (Figure 12.2). Through these decades, the new farms grew many other crops including corn, wheat, tobacco, rye, flax, peas, beans, potatoes, and buckwheat, as well as farming cattle and hogs, but cotton was clearly predominant.

In the Carolina Piedmont, roads were opened and river transport expanded in the early 19th century. Populations of Europeans and Africans grew rapidly, and in the southern Piedmont the percentage of the total population who were slaves increased from 20% in 1790, to 40% in 1820, to 60% in 1860, an increase directly tied to the growth of cotton.

Throughout this period, soil management and its improvement became a goal of many planters. Regional uniformity of cultivation methods increased as the 19th century progressed (Gray 1933). A typical agronomic regime included cotton seeds being sown heavily and thinned with a hoe in a process known as "chopping out". Chopping out left about one plant per foot by mid-July. The crop was weeded periodically but allowed to grow through the summer. Up to three harvests were made in the fall. After final harvest, dried stems and foliage were hand clubbed and plowed under, returning at least a small amount of organic matter to the soil.

Cotton pests were periodically rampant. These included root rots, anthracnose, sore-shin, blights, army worms, black rot, cutworms, cotton lice, ants, rust, and finally, in the early 20th century, the boll weevil. To eradicate pests, cotton was burned or rotated with other crops. Turkeys and chickens were occasionally run through cotton fields infested with insects.

By the middle of the 19th century, fertilizers had started to become popular, although they were not at all heavily applied. Fertilizer amendments included manure and compost, swamp mud, lime, plaster, ground bones, blood, wool, and guano (Sheridan 1979). These were the same decades in which Rothamsted Experimental Station in southern England was being organized, the time of Justis von Liebig's formulation of the law of the minimum, the period when E.W. Hilgard (1860), as State Geologist of Mississippi, rode across Mississippi on horseback collecting soil samples for chemical testing and land-use recommendations.

In response to the desire of planters to improve soil management, many agronomic methods were publicized. Planters' clubs and societies proliferated. Agricultural fairs and agricultural societies were initiated. Many farmers read journals such as *Southern Agriculturist*, *Southern Planter*,

Southern Cultivator and Dixie Farmer, Carolina Planter, Farmers Journal, and *American Farmer*. Popular books included Edmund Ruffin's (1852) *An Essay on Calcareous Manures*, John Taylor's (1813) *The Arator*, and John Binns' (1803) *Treatise on Practical Farming*. Plowing on the contour was advocated, as were crop rotations, runoff-diversion ditches, and terraces. The value of crop rotations was appreciated, despite (or perhaps because of) the great temptation to crop cotton continuously. Gray (1933) described continuous cotton as increasingly common, although cotton was also rotated with corn and sometimes cowpeas. Less common cotton rotations were cotton–corn–small grain and cotton–corn–fallow (Hilgard 1860).

Hypothetically, the most common soil fertility problems on Piedmont Ultisols were probably associated with nitrogen deficiencies and the complex of problems associated with acid-soil infertility. These included low bioavailable phosphorus and aluminum toxicity. Cotton is sensitive to low soil pH and aluminum toxicity, and it was likely that acidic B horizons restricted rooting and made cotton more sensitive to drought.

The popularity of lime and marl additions of Ruffin (1852) and crop rotations of Binns and Janey (Binns 1803) is intriguing. Based in Virginia, Binns recommended periodic rotations of clover with additions of plaster ($CaSO_4$). Applications of lime, marl, or plaster plus a cover of clover increased the bioavailability of calcium, sulfur, and nitrogen. The sulfate in gypsum relieved deficiencies of sulfur which in the early 19th century may well have been common. Deficiencies of soil sulfur are a rarity in the southeastern USA today, due mainly to inadvertent sulfur fertilization from use of superphosphate fertilizer (which contains sulfur) and to inputs of air-pollutant sulfur (Richter and Markewitz 1995a).

By the time of the USA's Civil War in the 1860s, soils in the southeast were pushed hard by agricultural practices that tended to deplete native fertility. Although many types of soils had come under widespread cultivation, upland Ultisols were most at risk of adverse management effects.

A hypothetical nitrogen budget for antebellum cotton (Table 12.2) demonstrates that without renewal of nitrogen, whether from N_2 fixation or from fertilization, the system leads to soil-nitrogen depletion and to productivity declines. Without nitrogen fertilization of antebellum cotton, productivity declines necessitated clearing of "fresh soil," and a system of shifting-field cultivation.

Table 12.2. *Hypothetical annual nitrogen budget for antebellum cotton in the uplands of the southern Piedmont, USA. Removals of nitrogen greatly outpace nitrogen inputs and the system is rapidly depleted of soil nitrogen*

Removal or input	Yearly nitrogen flux (kg ha^{-1})	Comments
Nitrogen removals		
Harvests	10–35	Removals in cotton harvests of seed plus lint at yields of 0.3 to 1 Mg ha^{-1} with 3.5% total N
Leaching	< 1–5	Drainage of 50 cm year^{-1} at < 0.2 to 1 mg L^{-1} nitrogen
Erosion	5–30	Erosional loss at 10 to 30 Mg ha^{-1} at 500 to 1000 μg g^{-1} of total N
Total losses	15–70	
Nitrogen inputs		
Atmospheric deposition	1–2	Pre-industrial nitrogen deposition
N$_2$ fixation	< 1	Free-living fixation; rotations with legumes only infrequently used
Fertilization	0	Not widely practiced
Total inputs	< 3	
Total system change	− 15 to − 70	Soil nitrogen deficit ranges up to 70 kg ha^{-1}

POST-WAR COTTON: MID-19TH TO 20TH CENTURY

After the incredibly destructive Civil War, i.e., post-1865, nearly all farmers in the southeastern USA, black and white, faced almost insurmountable operational problems. Communities were decimated. Nearly four million African-Americans had been freed from slavery, yet few owned land, animals, or farm implements. Credit and cash were limited or non-existent. Systems of communication and transportation were problematic at best. Agricultural statistics of the 1870 Census illustrate a region that was crippled, poor, and not likely to move ahead rapidly. It was a painful rebuilding of the agricultural economy from the scale of the individual farm such as the Old Ray Place (Table 6.2) to the entire region.

Remarkably, agricultural production did pick up. By 1880, cotton harvests from southern soils were equal to or greater than those in 1850 and 1860. In South Carolina, cotton production in 1880 was larger than it had been in 1860. And cotton production in 1890 was even greater

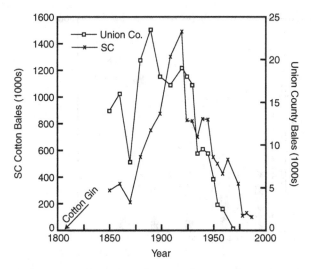

Figure 12.3. Estimated production of cotton from the state of South Carolina and from Union County, the county in which the Calhoun ecosystem is located (US Agricultural Census data, 1850–1987). Differences in weights of cotton bales have relatively small effects on the temporal patterns. Cotton production decreased in Union County prior to the more general state-wide decrease.

than in 1880 (Figure 12.3). For the region, and for South Carolina in particular, the Civil War appears now to have been a severe but temporary interruption in an upward trajectory of cotton harvests (Figure 12.3).

The recovery of cotton production following the Civil War was attributable to many causes: greater areas planted in cotton, introduction of fertilizers and lime, and more continuous cultivation. The recovery was so impressive that E.W. Hilgard, a "father" of soil science according to Hans Jenny (1961b) and the former State Geologist of Mississippi (Hilgard 1860), was commissioned by the US Department of Agriculture to organize an unprecedented two-volume study of United States cotton for the 1880 Agricultural Census.

In the final three decades of the 19th century, 200 million bales of cotton (approximately 90 billion pounds) were produced from southern soils, a production worth nearly $10 billion according to Cowdrey (1996). Although there was enormous local variation in cotton production in the post-war South, cotton had rapidly regained its national significance as a major agricultural commodity. As Tom Clark (1968) remarked about post-war conditions, "the cotton gin was as much a landmark in a southern town as the Confederate Monument."

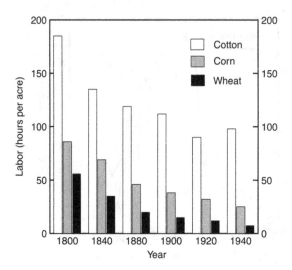

Figure 12.4. Estimated labor requirements of cotton, corn, and wheat (1800–1940), a pattern that illustrates the high labor demands of cotton at the same time as other crops were becoming highly mechanized (Fulmer 1950).

Down on the farm, however, cotton production was incredibly difficult. Cotton farming was labor-intensive, and on many farms not very profitable. Especially compared with other row crops such as wheat or corn, cotton demanded high labor inputs (Figure 12.4). Mechanical improvements for cotton production were relatively minor until well into the 20th century. Until the 1940s, for example, human labor, mules, and horses supplied most of the power to manage cotton fields (Fulmer 1950). Due to the enormous amount of hand labor needed to grow and harvest cotton, it had a greater bearing on human lives than any other crop grown in North America (Figure 12.5). Only by the 1940s was an effective mechanical cotton picker invented and marketed. By then, cotton was declining rapidly in the southern Piedmont.

Tenant farming and sharecropping became the common methods by which cotton was cultivated after the Civil War. Sharecroppers were supplied land, house, tools, and animals, often in exchange for half the crop (Vance 1929). Tenant farmers owned their own tools and work animals, but worked a landlord's land and lived in the owner's house, all in exchange for half to three-quarters of a crop. Cotton for tenant farmers and sharecroppers was often a trap from which there was little comfort, education, or likelihood of self-improvement. Living conditions on small family farms were often abysmal for blacks and whites (Vance 1929; Johnson et al. 1935).

(a)

(b)

Figure 12.5. Cotton production was an extremely difficult, labor-intensive occupation (Holley *et al.* 1940). (a) Cotton chopping in the early growing season, (b) one-half row cotton cultivation, and (c) *overleaf* one-row cotton cultivation, all from the 1930s.

Figure 12.5 (*cont.*)

Across the southeastern USA, cotton was the principal money crop on which merchants and bankers made loans. Lack of capital ensured that tenants and sharecroppers borrowed against the expectation of their next harvest. Many became obligated to grow cotton to obtain credit, a situation that certainly promoted continuous cultivation.

Although variation existed among fields and farms, the impact of cotton on soil was in general disastrous. Enormous numbers of farmers were destitute and the soil that they managed reflected their economic condition. Johnson *et al.* (1935) described the southern landscape as

> ... a miserable panorama of unpainted shacks, rain-gullied fields, straggling fences, rattle-trap Fords, dirt, poverty, disease, drudgery, and monotony that stretches for a thousand miles across the cotton belt.

Accelerated erosion occurred at phenomenal rates and created massive gullies. According to soil erosion maps of Trimble (1974), accelerated erosion was worst in the Piedmont counties of South Carolina and Georgia, including Union County and the Calhoun Experimental Forest.

Exposed red clay and gullies made a significant impact on visitors to the region and caught the attention of critics almost as much as did the plight of the sharecroppers and tenant farmers who worked in the

Figure 12.6. Many soils were not adequately protected when cropped to cotton and other row crops. As a result they were often severely eroded by the intense rainfall of the southeastern USA, sometimes to the point of gullying. The erosion damage to the soil was dramatic and graphic, most especially in the southern Piedmont (Trimble 1974). The photograph, by C. Korstian, dates from about 1930 and is from Duke Forest, Durham County, NC. It is not necessarily representative of soil damage done to the entire region, but it does illustrate the severity of erosion on many farms.

red-clay fields. In the 1930s, a popular perspective developed throughout the nation that the southern farmer was a "soil miner," who, it was said, "dug out the fertility from the soil, sifted it for profits, and deposited it in waste-spoils of degenerate old fields" (Earle 1992). To paraphrase a southern governor who spoke about the wasteful results of sharecropping (Vance 1929), "The cropper skins the land, as the landlord skins the cropper." Although these views of southern farmers are now criticized as being entirely too simplistic (Earle 1992), continuous cropping with inadequate management can be extremely damaging to soil. Without adequate vegetative protection and care for the soil, it was degraded by erosion (Figure 12.6), often spectacularly so (Hall 1948; Trimble 1974).

THE SOUTH'S INCREASING USE OF FERTILIZERS

From the early 1800s, when gentleman farmers of the Old South wrote about the beneficial effects of lime and manure, it has been widely known that crop production could be made more continuous with nutrient amendments that compensated for those removed in harvest. After the Civil War, applications of fertilizers slowly became a part of farming.

Even by the mid-20th century, however, agricultural scientists still complained about the disparity between extension agents' high rates of fertilizer recommendation and farmers' actual applications. A South Carolina Experiment Station pamphlet stated bluntly: "proper fertilization has not been followed by a sufficient number of farmers to promote noteworthy results" (Peterson and Aull 1945). The Tennessee Valley Authority, internationally known for its hydropower dams, regional electrification, and river-basin planning, also had a substantial impact on the southeastern region with its programs to promote fertilization. Federal control of acreage under major crops also encouraged higher applications of fertilizers.

During the last half of the 19th century, phosphorus fertilization grew rapidly. The first large-scale phosphate mines and production facilities were developed in coastal South Carolina in the 1870s (Sheridan 1979). In the late 19th century and throughout the 20th century, phosphate mining expanded in Florida, North Carolina, and Tennessee. These operations supported the rising demands for phosphorus in southern USA soils, and grew to satisfy a large fraction of the global market for phosphorus fertilizer as well. A number of these phosphorus mines are still active today.

The rising demand for nitrogen by southern USA agriculture was also instrumental in the development of the industrial capacity to supply fertilizer nitrogen. At the start of the 20th century, agricultural demand for fertilizer nitrogen was met by guano, fish scraps, dried blood, bone meal, and cottonseed meal. Organic sources supplied about 90% of the fertilizer nitrogen production in the USA in 1900 (Sheridan 1979). By 1910, nearly half of the USA's demand for fertilizer nitrogen was met by the extraction of nitrogen from coal. By 1920, however, the Haber–Bosch technology, developed to supply explosives for World War I, was redirected to nitrogen-fertilizer production. The Haber–Bosch process revolutionized nitrogen fertilization. The results are evidenced by rates of global nitrogen consumption illustrated in Figure 1.5.

Potassium used in southern agriculture has been largely imported. Near the end of the 19th century, German deposits were the primary sources, and Germany monopolized the world potassium-fertilizer market for several decades. After World War I, potassium-containing deposits in the western USA and in Canada helped meet demands of the southern market.

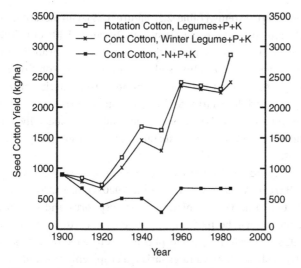

Figure 12.7. Cotton yields since 1900 from three soil treatments in the Old Rotation study, Alabama, USA (Mitchell 1988). The general decreases in cotton yields from 1900 to 1920 are attributed to the combined effects of phosphorus deficiencies and depredation by the boll weevil. Since the 1920s, phosphorus has been added in all treatments and the boll weevil brought under control. The soils without nitrogen additions continue to produce cotton at $> 500\,kg\,ha^{-1}$, whereas soils with nitrogen additions have steadily increased their production. Old Rotation is especially relevant to the Calhoun study since it tests soil effects of long-term cotton on the Pacolet series soil, an acidic, kaolinitic Ultisol, similar in several ways to the Appling soil at Calhoun.

EFFECTS OF COTTON AGRICULTURE ON SOIL FERTILITY: THE CASE OF OLD ROTATION

Applications of lime, nitrogen, phosphorus, and potassium became standard components of soil management for cotton, even by sharecroppers and tenant farmers. Such inputs of soil nutrients were required to maintain at least some yield and profit, especially if cotton was to be continuously cropped.

Cotton's long-term effect on soil fertility and crop productivity has been studied at Auburn University in Alabama, USA since the 1890s (Mitchell *et al.* 1991, 1996). The best-known study is Old Rotation. This study demonstrates clearly how soil management that includes crop rotations and nutrient management can compensate for harvest removals from crop systems (Table 2.2, Figure 12.7). Old Rotation is especially relevant to the Calhoun ecosystem, because fields of Old Rotation contain Pacolet soils: acidic Ultisols derived from granitic gneiss and close relatives of the Appling soils of Calhoun. Old Rotation is the world's longest-running study that examines sustainability of a crop system on a soil with low native fertility, i.e., on an Ultisol.

Three outcomes of the Old Rotation study are especially significant to understanding how soils sustain their nutrient supply under management.

The first outcome is that cotton yields in Old Rotation plots decreased in all treatments between 1900 and 1920 (Figure 12.7). Decreases were attributed to harvests depleting soil phosphorus and to depredation by the boll weevil, both common problems in cotton grown in that era. After 1924, yearly fertilizer phosphorus was boosted from 11 kg ha^{-1} to >20 kg ha^{-1}. These elevated rates satisfied crop demands, and in 1988 available phosphorus in surface soils was estimated by Mehlich I extraction to be between 40 and 50 μg g^{-1}, a very high index of bioavailability. After 64 years of phosphorus fertilization at 20 to 45 kg ha^{-1} year^{-1}, these Ultisols could probably go without phosphorus fertilization for several rotations with little decline in productivity.

The second outcome is that without nitrogen inputs cotton productivity diminished substantially. Cotton yields decreased to about 0.5 Mg ha^{-1} year^{-1} in "minus-N treatments," plots with no nitrogen fertilizer added or leguminous crops grown (Figure 12.7). After a century of cropping without added nitrogen, soil-nitrogen inputs and removals are approaching an equilibrium in the minus-N plots. In these plots, nitrogen contained in cotton harvests amounts to only 13 kg ha^{-1} year^{-1}, a removal rate that is supplied from atmospheric nitrogen deposition (estimated at 5 to 10 kg ha^{-1}, Richter and Markewitz 1995a), plus meager amounts from N$_2$ fixation by free-living microbes (probably <1 kg ha^{-1}) and a small amount of net nitrogen mineralization from continued decomposition of soil organic matter. In the minus-N plots, soil organic nitrogen has been greatly depleted by long-term cropping without fertilizer nitrogen amendments or leguminous cover crops. The nitrogen cycle of the minus-N plots of Old Rotation (Table 2.2) is comparable with the nitrogen cycle of the antebellum management of cotton detailed in Table 12.2.

The third outcome is that rotations of cotton with winter legumes, vetch, and crimson clover benefit cotton productivity as much as fertilization with inorganic nitrogen added at 130 kg ha^{-1} year^{-1}. Well managed leguminous cover crops fix large amounts of nitrogen, enough to benefit crops and soil organic nitrogen and to maintain high levels of productivity, at least in this system in Alabama, USA.

Results from Old Rotation suggest a pattern of effects that cotton management had on soil productivity and fertility, including the adverse impacts that cotton had on soil without adequate management (Table 2.2). The next chapter examines specific effects of Old South agriculture on soil biogeochemistry.

13

Agricultural legacies in old-field soils

134   The full details of how agriculture has altered soils in the southeastern USA are difficult to determine, since almost all of the landscape has been cultivated and substantially affected by agricultural management for up to three centuries.

In Chapters 8 and 9, soils of hardwood forests in the Calhoun vicinity were examined. These soils were judged to be potentially arable, but had remained uncultivated and unfertilized (Table 9.1). In this chapter, we contrast these uncultivated soils under hardwoods with 12 long-cultivated versions (Table 13.1).

All 12 cultivated soils were probably first cleared of their primary forest and cultivated for cotton and other crops early in the 1800s. Five of these 12 cultivated soils were old cotton fields that during the mid-20th century were converted from cotton to hayfields and have continued to receive fertilizers, lime, and other management inputs. Six others of the 12 cultivated soils were old cotton fields that during the mid-20th century were planted with loblolly pine seedlings. These old-field soils under pine have not been fertilized or limed since they were last in agriculture. One soil of the 12 is still cultivated and fertilized (Table 13.1).

All 17 soils lie on interfluves with $< 10\%$ slopes (Table 13.1) and have coarse-textured A horizons, that are sandy loams or loamy sands. All previously cultivated soils are moderately but not seriously eroded, especially by regional standards. Our critical assumption for this comparative study is that all 17 ecosystems had similar Ultisol soils derived from granitic gneiss prior to 12 of them being converted to agriculture in the early 1800s.

This is a comparative soil study, not a chronosequence, as much more than Jenny's (1980) "age t" has affected these soils that currently support hardwoods, agro-ecosystems, and old-field pines. Although soils

Table 13.1. *Some details of the four ecosystems from which granitic gneiss-derived soils were sampled to investigate effects of long-term agriculture on soil properties (Dunscomb 1992). The soil-comparative study has a randomized incomplete block design. Bulk density is the mean of five samples per site*

Ecosystem	Site code	Site name	Soil series	Approx-imate slope (%)	Bulk density, 0–0.075 m (Mg m^{-3})
Oak–hickory[a,b,c]	H-1	Calhoun	Appling	2	1.17
Oak–hickory[a,b]	H-2	Murphy	Madison	10	0.93
Oak–hickory	H-3	Padgett's Creek Church	Cecil	8	—
Oak–hickory[a,b]	H-4	Mt. Zion Church	Cecil	2	1.06
Oak–hickory[a,b]	H-5	Rt. 196	Cecil	5	1.10
Loblolly pine[c]	P-1	Calhoun	Appling	3	1.30
Loblolly pine[a,b]	P-2	Murphy	Madison	5	1.14
Loblolly pine[a,b]	P-3	Padgett's Creek Church	Cecil	4	1.24
Loblolly pine[b]	P-4	Mt. Zion Church	Cecil	2	1.14
Loblolly pine[a,b]	P-5	Rt. 196	Cecil	5	1.19
Loblolly pine[a,b]	P-6	Greer	Appling	3	1.28
Hay[a,b,c]	G-1	Calhoun	Appling	2	1.32
Hay[a]	G-2	Murphy	Cecil	5	1.29
Hay[a,b]	G-3	Padgett's Creek Church	Cecil	5	1.06
Hay[a,b]	G-4	Mt. Zion Church	Cecil	3	1.28
Hay[a,b]	G-6	Greer	Appling	3	1.40
Row crop[a]	C-1	Calhoun	Appling	2	1.38

[a]Sites with intensive 0 to 0.3 m sampling.
[b]Sites with intensive 0 to 1.1 m sampling.
[c]Sites with 0 to 5 m sampling.

of the hardwood forests have soil profiles that most closely correspond to conditions prior to about 1800 when the region was initially cleared and cultivated, the hardwood forests are not perfect representations of pre-1800 conditions. Hardwood forests after all have served farmsteads as grazing areas and woodlots for fuelwood and timber, and have received modern atmospheric-deposition inputs of nitrogen, sulfur, and nutrient cations.

The 12 cultivated soils also have a somewhat complicated short-term pedogenesis. The soils that continue to be agriculturally managed

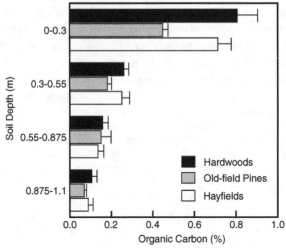

Figure 13.1. Concentration of soil organic carbon in the upper soil profile of uncultivated hardwoods, old-field pines, and currently managed hayfields in or near the Calhoun Forest Experiment, SC. Most obvious effects of land use on soil carbon are in the upper 0.3 m. Means and standard errors are illustrated.

probably receive higher rates of fertilizers and lime than previously cultivated soils that currently support old-field pines that are 40 to 50 years old. In other words, fertilizer and lime rates today are higher than they were 40 to 50 years ago when the soils under old-field pine were last managed for cotton. Despite these caveats, the comparison of the three ecosystems' soils is highly informative mainly because of the major soil transformations that have been affected by agriculture across the region.

In this chapter, the comparative soil study demonstrates how long-term cotton substantially depleted soil organic matter and organic carbon, and ameliorated soil acidity, base saturation, organic nitrogen, and bioavailable phosphorus in soils not severely eroded in the Calhoun Experimental Forest. Not only were these soil properties substantially altered by past agricultural management, but they appear to remain significantly altered for many decades and perhaps centuries after agricultural abandonment.

ORGANIC CARBON

Organic carbon in the upper 0.3 m of mineral soil of unfertilized and uncultivated hardwood forests averages 0.80% (CV% = 24.4 among four hardwood forests) (Figure 13.1). A distinct A horizon is present in these uncultivated soils. Although annual crops are no longer an important component of land uses in this part of the 21st century Piedmont, we estimate that past long-term cultivation for cotton, corn, and other crops reduced organic carbon in the upper 0.3 m of soil to about 0.5%. Soil in the one row-crop field in this study averages 0.53% carbon in this surficial layer. Similarly, the upper 0.3 m of mineral soil under old-field pines (cultivated until about 40 to 50 years ago) averages 0.45% carbon (CV% = 12.5 among five pine stands), suggesting a major reduction in soil carbon as well as a minimal rate of carbon reaccumulation under pines. Under pine, only an incipient A horizon can be observed. Under fertilized hayfields, however, soil organic carbon averages 0.71% (CV% = 20.2 among five fields), indicating that the fertilized and limed grass-dominated fields have elevated soil organic carbon. An A horizon is generally obvious under these hayfields.

Cultivation reduces soil organic matter by both reducing inputs of plant debris and enhancing decomposition due to elevated soil tempera-ture and physical mixing from plowing. If soils under pines are re-accumulating soil organic matter, they appear to be doing so at relative-ly low rates. Why pines apparently accumulate soil carbon at relatively low rates cannot be addressed in this comparative soil study, but is a question directly addressed by the repeated sampling of the long-term Calhoun experiment described in Chapter 15 (Richter et al. 1999).

Agricultural practices increased bulk density (mass of soil per unit volume), especially in the upper 0.15 m of soil (Tables 13.1, 13.2). Because of the greater density of cultivated soils, effects of agriculture on soil carbon were greater on carbon concentration scaled to a unit-weight basis (carbon in % by weight) than on soil-carbon storage scaled to a unit-area or unit-volume basis (carbon content in $Mg\ ha^{-1}$ or $kg\ L^{-1}$). Carbon storage under the uncultivated hardwood forests totals 32.5 Mg ha^{-1}, whereas under old-field pine and row crops it amounts to 19.1 and 23.5 $Mg\ ha^{-1}$, respectively, in the upper 0.3 m of mineral soil (Table 13.2). Soil-carbon concentration and content under old-field pine or row crops appears to be about 60% of those under hardwoods, suggesting that cultivation reduced soil carbon by about 40% of that present in the upper 0.3 m depth prior to cultivation. By contrast, soil (0 to 0.3 m) under hay contains similar amounts of carbon to that under hardwood forests, about 32.0 $Mg\ ha^{-1}$. We attribute this relatively high mineral-soil carbon to elevated carbon inputs both above- and belowground that have been stimulated by fertilization and lime, lower decomposition under grass, and at least a slight reduction in soil temperatures compared with clean cultivation.

Table 13.2. *Mean carbon and bulk densities of surficial 0.3 m layers of soil from the four ecosystems in or near the Calhoun Experimental Forest, SC (Dunscomb 1992). Coefficients of variation are in parentheses. Collections were made in 1992*

Depth (m)	Hardwood ($n = 4$)	Old-field pine ($n = 4$)	Hayfield ($n = 5$)	Row crop ($n = 1$)
Total carbon (%)				
0–0.075	1.927 (23.2)	0.747 (35.4)	1.475 (19.8)	0.814
0.075–0.15	0.808 (44.1)	0.446 (19.8)	0.577 (29.4)	0.698
0.15–0.3	0.394 (7.2)	0.336 (27.6)	0.479 (44.1)	0.327
0–0.3[a]	0.805 (24.4)	0.449 (12.5)	0.712 (20.2)	0.529
Total carbon (Mg ha^{-1})				
0–0.075	15.73 (29.1)	6.67 (30.5)	14.01 (22.1)	7.69
0.075–0.15	8.08 (40.2)	4.86 (20.6)	6.67 (26.7)	7.94
0.15–0.3	8.70 (6.6)	7.54 (24.6)	11.31 (42.6)	7.90
0–0.3	32.52 (24.4)	19.07 (12.5)	31.99 (20.2)	23.53
Bulk density (Mg m^{-3})				
0–0.075	1.08 (10.7)	1.20 (5.0)	1.27 (10.1)	1.38
0.075–0.15	1.36 (6.6)	1.45 (3.8)	1.56 (9.2)	1.52
0.15–0.3	1.47 (2.1)	1.50 (4.4)	1.58 (4.8)	1.61
0–0.3	1.35 (3.6)	1.41 (3.9)	1.50 (5.6)	1.53

[a]Average weighted by depth and bulk density.

ORGANIC NITROGEN AND C/N RATIOS

The dynamics of soil nitrogen contrast with those of soil carbon (Figure 13.2), as nitrogen content is relatively high in soils managed for agriculture, due no doubt to long-term nitrogen fertilization. In currently fertilized row crops and hayfields, for example, total nitrogen in the upper 0.3 m of soil averages 396 and 520 µg g^{-1}, or 1765 and 2337 kg ha^{-1}, respectively (Table 13.3). Total nitrogen in the unfertilized and uncultivated soils that support hardwood forests averages 384 µg g^{-1} (CV% = 36.4) or 1551 kg ha^{-1} in this surficial soil depth (Table 13.3). In old-field pines, however, there are markedly lower concentrations and contents of soil nitrogen: 238 µg g^{-1} (CV% = 23.8) and 1010 kg ha^{-1}. The relatively low soil nitrogen under pine is attributed to a combination of rapid nitrogen accumulation by vigorously growing forests (that have transferred soil nitrogen to tree biomass and forest floor) and lower nitrogen fertilization rates 40 to 50 years ago compared with current inputs to agricultural soils. Even still, much of the nitrogen in the

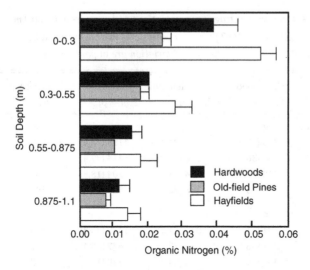

Figure 13.2. Concentration of soil organic nitrogen in the upper soil profile of uncultivated hardwoods, old-field pines, and currently managed hayfields in or near the Calhoun Forest Experiment, SC. Organic nitrogen appears elevated throughout the upper meter of soil, presumably an effect of immobilization of fertilizer N. Means and standard errors are illustrated.

current pine stand's nitrogen cycle appears to be derived from agricultural fertilizer added to soils in the distant past. The soil comparison suggests that the nitrogen cycle changed substantially in pine-forest and hayfield ecosystems during the 20th century.

Atmospheric deposition of nitrogen to this region is estimated at 5 to 10 kg ha^{-1} year^{-1} (Richter and Markewitz 1995a). Cumulatively, such nitrogen inputs can be biologically significant on the time scale of decades, especially in soils and ecosystems that retain inputs. Over 50 years, for example, yearly nitrogen deposition of 5 to 10 kg ha^{-1} (Richter and Markewitz 1995a) has totaled 250 to 500 kg ha^{-1}, an important fraction of these ecosystems' total nitrogen capital (Table 13.3).

The dynamics of soil nitrogen and soil organic matter can be illustrated by the C/N ratio, an index of organic matter quality that has been substantially altered by agricultural use (Table 13.3, Figure 13.3). Soils supporting current row crops and hayfields have relatively low C/N ratios of 13.4 and 13.7 (CV% = 17.4), respectively. The C/N ratios were reduced both by nitrogen fertilizer's enrichment of soil organic nitrogen and by cultivation's stimulation of organic carbon decomposition. In the surficial 0.3 m of soil, the C/N ratios of soils under forests are considerably higher than under agricultural crops (Table 13.3). The hardwoods

Table 13.3. *Mean total nitrogen concentrations and contents of 0 to 0.3 m mineral soil collected in 1992 from four ecosystems in or near the Calhoun Experimental Forest, SC. Coefficients of variation are in parentheses*

Depth (m)	Hardwood ($n = 4$)	Old-field pine ($n = 4$)	Hayfield ($n = 5$)	Row crop ($n = 1$)
Total nitrogen ($\mu g\ g^{-1}$)				
0–0.075	853 (37.6)	346 (22.1)	1198 (23.6)	550
0.075–0.15	363 (36.3)	200 (23.2)	414 (37.4)	500
0.15–0.3	218 (49.1)	214 (41.6)	304 (18.8)	280
0–0.3	384 (36.4)	238 (23.8)	520 (19.5)	396
Total nitrogen ($kg\ ha^{-1}$)				
0–0.075	701 (41.0)	312 (11.3)	1139 (25.8)	519
0.075–0.15	367 (36.4)	218 (27.1)	477 (35.4)	569
0.15–0.3	483 (50.3)	480 (44.9)	720 (18.7)	677
0–0.3	1551 (36.4)	1010 (23.8)	2337 (19.5)	1765
C/N ratio				
0–0.075	22.6 (28.2)	21.5 (24.4)	12.3 (10.5)	14.8
0.075–0.15	22.3 (29.2)	22.7 (16.9)	14.0 (22.8)	14.0
0.15–0.3	18.1 (69.0)	17.3 (30.1)	15.8 (29.9)	11.7
0–0.3	21.0 (37.2)	19.6 (22.0)	13.7 (17.4)	13.4

C/N ratio averages 21.0 (CV% = 37.2), and that of the old-field pine averages 19.6 (CV% = 22.0). Assuming that the C/N ratio of soils under old-field pines was once reduced by fertilization, pine reforestation appears to have increased C/N ratios.

In Chapters 15 and 16 the rate and processes of carbon and nitrogen change in soils and ecosystems are directly evaluated in repeated samplings during the 40-year development of the old-field Calhoun pine forest.

SOIL PHOSPHORUS

Phosphorus fertilization has had substantial and remarkably long-term effects in elevating soil-extractable phosphate, particularly in surficial layers of soil. Previously fertilized soils have phosphorus extractable in dilute hydrofluoric and nitric acid (Mehlich III extraction, Page 1982) that averages $>6\ \mu g\ g^{-1}$ in the upper 0.3 m of soil, whether these are currently fertilized soils in hayfields or soils under pine last fertilized 40 to 50 years ago (Figure 13.4). The dilute-acid-extractable phosphorus in fertilized soils averages more than 3-fold the phosphorus extractable in

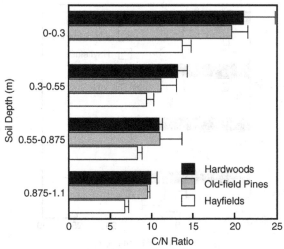

Figure 13.3. The ratio of C/N in the upper soil profile of uncultivated hardwoods, old-field pines, and currently managed hayfields in or near the Calhoun Forest Experiment, SC. The C/N ratio differences among land uses are relatively large. Means and standard errors are illustrated.

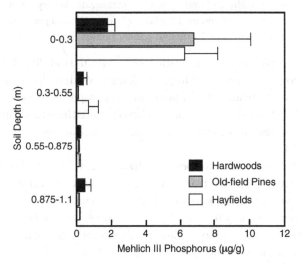

Figure 13.4. Soil extractable phosphorus (Mehlich III) concentration in the upper soil profile of uncultivated hardwoods, old-field pines, and currently managed hayfields in or near the Calhoun Forest Experiment, SC. Means and standard errors are illustrated.

soils under hardwoods (Figure 13.4). These Ultisols with chemically active hydrous Al- and Fe-oxides surfaces strongly adsorb fertilizer phosphorus and a fraction of this phosphorus appears to remain extractable for many decades.

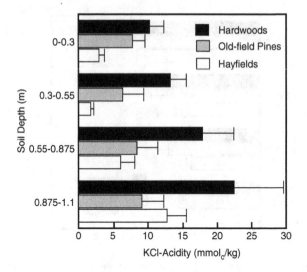

Figure 13.5. KCl-acidity concentration in the upper soil profile of uncultivated hardwoods, old-field pines, and currently managed hayfields in or near the Calhoun Forest Experiment, SC. Unlimed soils (under hardwood) are extremely acidic, and liming reduces exchangeable acidity throughout at least a meter of soil and has long-lasting effects. Old-field pine soils were limed 40 to 50 years ago. Means and standard errors are illustrated.

Without phosphorus fertilization, the acidic, upland Ultisols have very low phosphorus bioavailability. According to Wells *et al.* (1973), soil phosphorus below which loblolly pine exhibits phosphorus deficiency is 3.0 μg g^{-1} as extracted by dilute acid solutions. Unfertilized soils under hardwoods average less than 1.8 μg g^{-1} (CV% = 47.7 among four hardwood stands).

Whether the soils have been fertilized or not, bioavailability of phosphorus in >0.3 m soil is low in all ecosystems. Concentrations of the acid-extractable phosphorus in all three systems between 0.3 and 1.1 m soil depth are less than 0.7 μg g^{-1}. Increases in extractable phosphorus that are attributed to previous fertilizer inputs do not appear to have penetrated beyond the upper 0.3 m (Figure 13.4).

In Chapter 18, we examine how the four-decade growth of the Calhoun old-field forest has altered several soil-phosphorus fractions, not only that extractable by dilute acids.

EXCHANGEABLE ACIDITY, BASE SATURATION, AND SOIL PH

One of the sharpest contrasts among the soils with different land-use histories is that of soil acidity. Soil chemistry has been greatly affected by

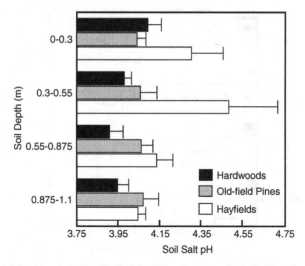

Figure 13.6. Soil pH in salt (KCl) in the upper soil profile of uncultivated hardwoods, old-field pines, and currently managed hayfields in or near the Calhoun Forest Experiment, SC. Means and standard errors are illustrated.

previous liming. Comparisons of soil under previously limed, old-field pine with unlimed hardwoods suggest that even 40 to 50 years after lime applications soils have diminished exchangeable acidity (Figure 13.5), elevated pH (Figure 13.6), and increased base saturation. Acidity has been partially neutralized by previous liming throughout the soil's upper meter, in A, E, and upper B horizons. The data also emphasize the extreme acidity of these soils in their unlimed condition (Figures 13.5, 13.6, Table 9.1).

In the surficial soil layers of the old-field pine stands, acidity appears to have reaccumulated during the 40 to 50 years of forest regrowth, especially in surficial layers. Because all of these Ultisols are dominated by low-activity kaolinite clays that have relatively small surface charge, they are not well buffered against changes in acid–base chemistry. Effective cation exchange capacity ranges between only 20 and 30 $mmol_c\,kg^{-1}$ in these soils, so we might expect relatively rapid changes in acid–base properties. Reforestation combined with the cessation of liming facilitates the readjustment of these Ultisols to their former more acidic state.

EXCHANGEABLE NUTRIENT CATIONS

Exchangeable calcium is remarkably low in concentration in the unlimed hardwood soils (Figure 13.7), averaging about 1.4 $mmol\,kg^{-1}$ in the

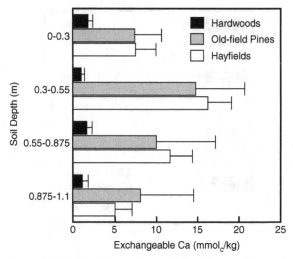

Figure 13.7. Soil exchangeable calcium concentration in the upper soil profile of uncultivated hardwoods, old-field pines, and currently managed hayfields in or near the Calhoun Forest Experiment, SC. Means and standard errors are illustrated.

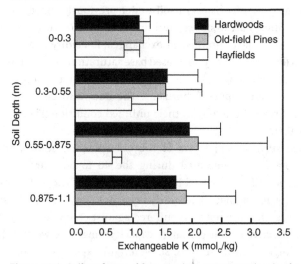

Figure 13.8. Soil exchangeable potassium concentration in the upper soil profile of uncultivated hardwoods, old-field pines, and currently managed hayfields in or near the Calhoun Forest Experiment, SC. Means and standard errors are illustrated.

upper 1 m of soil. In contrast are soils of the old-field pine stands and the currently limed hayfields, in which concentrations of exchangeable calcium are on average 7-fold higher than those under hardwood forests.

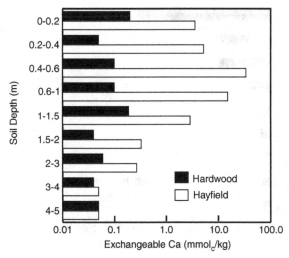

Figure 13.9. Soil exchangeable calcium concentration in the upper 5 m of an Appling series profile under unfertilized and unlimed hardwood and intensively managed hayfield (sites H-1 and G-1, respectively).

Although exchangeable magnesium was not apparently affected by land-use history, exchangeable potassium is less in hayfield soils than in either old-field pine or hardwood forests (Figure 13.8). Multiple harvests of hay each year may exert a high demand on the exchangeable potassium present in the soil.

HOW DEEPLY HAVE AGRICULTURAL EFFECTS PENETRATED PIEDMONT SOILS?

These soil-comparison data illustrate how long-term agricultural practices have substantially affected soil fertility, and that some effects have persisted for decades since last cropping.

Components most deeply affected in the profile by agricultural use are soil nitrogen, exchangeable calcium, base saturation, and soil pH; much less deeply affected are soil phosphorus and carbon. Two directly adjacent soils are compared in the Calhoun ecosystem (Figure 6.5): one supports an old, unfertilized hardwood forest (H-1) and the other was an old-cotton field that since the mid-20th century has been intensively managed for hay (G-1). Data presented from soils sampled down to 5 m depth come from cores manually augered or hydraulically sampled by a truck-mounted Geoprobe (Figure 9.4).

Soil exchangeable calcium appears to be deeply affected by long-continued agriculture (Figure 13.9). Although calcium supplied in lime

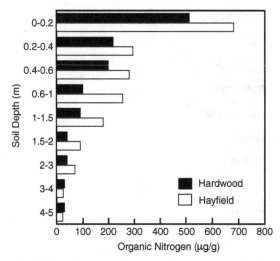

Figure 13.10. Soil organic nitrogen concentration in the upper 5 m of an Appling series profile under unfertilized and unlimed hardwood and intensively managed hayfield (sites H-1 and G-1, respectively).

is readily soluble in the short term, exchangeable calcium appears to be elevated throughout the upper 2 to 3 m of soil under currently limed hay compared with the soil under hardwoods. Even in soil between 1.5 and 2 m depth, exchangeable calcium is nearly 10-fold greater than that in the unfertilized hardwood soil.

Base saturation is similarly increased throughout the upper 1 to 1.5 m of soils that have been cultivated and fertilized. Soil pH is also increased in these profiles, although the depth of effect appears to be relatively less than the increases in base saturation and exchangeable Ca. In this comparison, soil pH measured in dilute $CaCl_2$ is elevated to at least 1 m depth, with a pH of 3.8 in 0.01 M $CaCl_2$ at 0.8 m depth under hardwoods and 4.2 under the limed hayfields.

Phosphorus is the fertilizer nutrient that has penetrated the soil profile the least of all fertilizer nutrients. As in the well replicated soil-comparison study, effects of fertilizer phosphorus have not penetrated below the plow zone of 0.15 to 0.2 m, despite inputs of phosphorus that likely have occurred over many decades.

As with phosphorus, differences in soil carbon are largely confined to surficial layers, the upper 0.15 m of mineral soil, with little indication of differences below. On the other hand, soil nitrogen appears to be enriched for >1 m depth in the fertilized hayfield ecosystems. Hypothetically, nitrogen from fertilizers leached and was incorporated into deep soil organic matter (Figure 13.10), reducing soil C/N ratios within the upper 2 m.

Not only has agriculture affected the soil environment to a depth that is impressive, but these effects may well persist long into the future. From a broad perspective, the agricultural legacy contained by soils in the southern Piedmont represents a profound transformation of these soils and ecosystems. A failure to recognize the degree and extent to which agriculture has transformed soils in the southern USA can produce, and no doubt has produced, misleading conclusions about the ecological functioning of the contemporary southern forest ecosystem.

Part IV

Soil change over time scales of decades: conversion of agricultural fields to secondary forests

"How do you know what you think you know?"

Prof. C.W. Ralston's epistemological question to graduate students speculating about effects of land management on soil

14

The birth of a new forest

Cotton production in the southern Piedmont reached its peak in the first two decades of the 20th century (1900–1920). Ironically, during this same time, small-farm economies became increasingly untenable (Healy 1985). During the 1920s, a wave of farm abandonment began in earnest. The economics of cotton made farming in the southern Piedmont unable to compete with larger, more efficient operations in other parts of the USA. During the economic depression of the 1930s, abandonment of Piedmont farms accelerated.

South Carolina agricultural statistics illustrate these land-use changes well. In 1920, cotton cultivation reached its greatest areal extent in the state with a total of 1.06 million hectares harvested (out of a total state area of 8.06 million hectares). By 1940, 478 000 hectares of cotton remained. By 1960, cotton was harvested on 220 000 hectares, and by 1982, only 38 400. The collapse of cotton in South Carolina was similar to the pattern of cultivated crops throughout the southern Piedmont; its abandonment affected nearly every aspect of the regional ecology, economics, and culture.

For three hundred years in eastern North America, "improved land" had been synonymous with agricultural land cleared of forest. But by the mid-20th century, farms had been extensively abandoned and secondary forests had established themselves on many millions of hectares of formerly cultivated soils. Environmental historian Williams (1989) wrote:

> In a society imbued with the frontier ideals of development, progress, and the virtues of forest clearing, abandonment was retrogressive, difficult to comprehend ... It was something to be ignored ...

Figure 14.1. Planting of forest on abandoned old fields of the Duke Forest, Orange County, NC in the 1930s (photograph C. Korstian).

Ecologists and foresters, however, did not ignore farm abandonment; they quickly became fascinated with how plant vegetation responded to this landscape being abandoned by farmers. In fact, the entire concept of plant succession was greatly enriched by classical studies of old fields in the central Carolina Piedmont by Billings (1938), Coile (1940), and Oosting (1942). Long-term vegetation studies that originated in the 1930s continue to be useful in examining plant-species dynamics in the aftermath of Piedmont-farm abandonment (Christensen 1989).

Not all cultivated fields that were abandoned were reforested, but large numbers were. Natural forest regeneration on most previously cultivated fields was dominated by southern pines including loblolly, shortleaf, Virginia (*Pinus virginiana*), longleaf (*P. palustris*), and slash (*P. elliotii*) pine. The pine stands often grew vigorously, even on seriously eroded soils. Within a decade after field abandonment, pines blanketed previously bare soil with needles and a new forest floor, and anchored the soil with extensive root systems. Slopes were stabilized and erosion substantially curtailed. Southern pines proved well suited for physically stabilizing abandoned farmland.

Tree seedlings were also planted on much of the former farmland, as a way to more closely manage reforestation and control agriculturally accelerated soil erosion (Figure 14.1). State and federal governments subsidized a number of reforestation programs to stimulate tree planting. Prior to the 1950s, governmental programs promoted planting of pine seedlings to control soil erosion and improve watershed management. After the 1950s, private land owners were encouraged to retire land from agricultural use by planting trees in the massive Soil Bank program. Although these were national programs, they were specifically

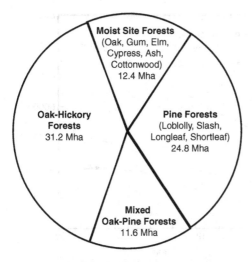

Figure 14.2. Areas of major forest types as determined by the US Forest Service in 1992 for the southeastern USA (13 states). Nearly 95% of the region's forests are included in these five forest ecosystems (Powell *et al.* 1993).

designed to gradually retire southern soils from further agricultural crop production.

A NEW FOREST, A NEW MANAGEMENT

The 20th century forest of the southeastern USA is a new type of forest. These new forest ecosystems share a legacy both with the land's original primary forest ecosystem, and with the land's agricultural past which has so extensively transformed the region's ecosystems biologically, chemically, and physically.

The present-day southeastern forest covers about 85 million hectares or nearly two-thirds of the total area of the 13-state southeastern region (Powell *et al.* 1993) (Figure 14.2). The forest is almost entirely secondary, in other words, regrown after being cut over, and in aggregate it currently grows at an incredibly high rate of net primary productivity. The modern southeastern forest possesses an economic and environmental value that is beyond all previous expectations.

The regrowth of the southeastern forest has greatly diminished agriculturally accelerated soil erosion in rural areas of the region. The much greater physical stability of forests compared with farmland is well documented at the scale of the individual farm (Tennessee Valley

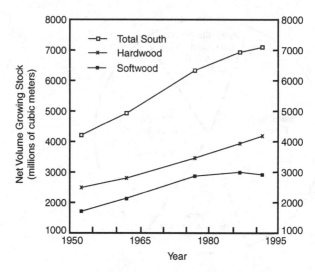

Figure 14.3. Changes in net volume of wood (i.e., growing stock) in south-eastern forests from 1952 to 1992 (Powell *et al.* 1993). Many decades of enormous accretion in the region's wood volume are coming to an end.

Authority 1962; Swank and Crossley 1988), but also at the scale of the river basin (Meade and Trimble 1974; Trimble 1974; Richter *et al.* 1995a). Even carefully practiced cultivation erodes soil at 10- to >1000-fold greater rates compared with forests.

Between the early 1950s and the 1990s, the southeastern forest nearly doubled its standing volume of wood (Figure 14.3), with net annual growth rates that were 3 to 5% of the standing wood volume region-wide (Figure 14.4). Such growth rates have been followed by increases in wood harvests for industrial products such as furniture, construction lumber, veneer, scaffolding, plywood, poles, pallets, and paper (Figures 14.5, 14.6). In the 1990s, far more wood was harvested for industrial-wood products from the forests in the southeastern USA than from any other wood-producing region in the world (Figure 5.3).

Remarkably, much of the soil resource that supports this prolific production of wood fiber supported cotton, tobacco, and corn in the past. Although this new level of forest growth directly impacts soils (and we examine these effects in upcoming chapters), it is important to call attention to the much less severe physical impact of modern forest management's periodic cycles of activity as compared with agriculture's annual disturbance cycles (Tennessee Valley Authority 1962; Trimble 1974; Swank and Crossley 1988; Richter *et al.* 1995a).

Modern forest management is not at all static in its approach to

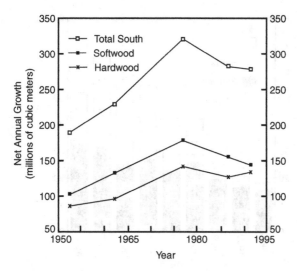

Figure 14.4. Net annual growth of forest wood volume in the southeastern USA (13 states). Net annual growth does not include losses due to mortality and damage from disease, insects, and fire (Powell *et al.* 1993).

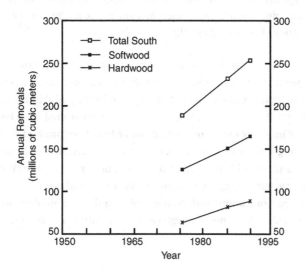

Figure 14.5. Annual removals of wood by harvest or land clearing in 13 southeastern USA states (Powell *et al.* 1993).

the forest ecosystem, but in aggregate is making a transition that is not dissimilar to that made by agriculturists when food gathering was transformed into cultivation agriculture. Modern forest management, especially in the southeastern USA, is becoming much more than an

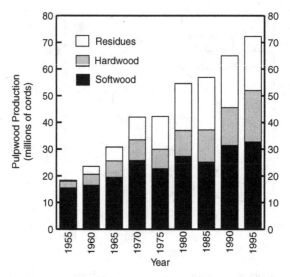

Figure 14.6. Annual production of wood for pulp from the 13-state south-eastern forest. Pine fiber continues to make up the major source for pulp, although hardwood fiber (mainly *Quercus* spp.) and residues from sawmill and pulpmill manufacturing each account for about 30% of the total production of wood used as pulp.

exercise in wood gathering. Modern silviculture is in the process of moving beyond the boom-and-bust bonanzas of timber extraction that too often have characterized logging enterprises in the past (Figure 14.7).

In the southeastern USA, forestry is being transformed from being an extractive industry to one interested not only in fiber production but also in the long-term management of forest ecosystems. This evolution is important to this book's objectives because a better understanding of soil and ecosystem change will almost certainly contribute to more intelligent decisions about soil management and water quality, wood production, and even the management of wildlife, aesthetics, and recreation.

MODERN SOILS IN THE SOUTHEASTERN USA FOREST

The enormous harvest of wood from the modern southeastern forest (Figures 5.3, 14.5) is based on a management that is being applied with an intensity and at a geographic scale never before attempted, on tens of millions of hectares of forested land. Questions about the sustainability of this intensively managed southern forest are far from trivial. Most of the upland soils in the region have low inherent fertility and, as we have

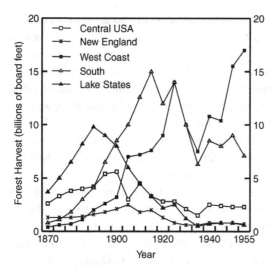

Figure 14.7. Boom-and-bust bonanzas of timber extraction from different regions of the USA (Williams 1989).

documented in previous chapters, fertility of the region's soils has been substantially impacted by agriculture. During the 1980s and 1990s, forest fertilization rapidly expanded across much of the southeastern forest, and in the late 1990s about a half-million hectares of pine forest were fertilized per year. Nearly all of these amendments have been nitrogen and phosphorus, although potassium and boron appear to be limiting nutrients in some soils.

Reforestation is generally considered to benefit formerly culti-vated soils; but rapidly growing plants, whether food crops or forest trees, place nutritional demands on soils to supply nitrogen, phos-phorus, potassium, calcium, magnesium, and micronutrients. In prin-ciple, the effects of rapidly growing secondary forests on soil nutrients are not dissimilar from those of food-crop systems in which plant biomass accumulates relatively large amounts of mineral-soil nutrients. Observations of how tree growth alters soil fertility over periods of decades are notably absent.

OPPORTUNITIES AND OBJECTIVES OF THE CALHOUN FOREST EXPERIMENT

The long-term Calhoun Forest Experiment presents major scientific opportunities due to its rigorous experimental design, its details of operation, and its continuity of record. The experiment has 16 permanent plots, which were installed in two fields long cultivated for cotton. Following the planting of loblolly pine seedlings in the winter of 1956–1957, soils have been sampled seven times over four decades of pine-forest development (1957–1997), about once every five years. Most valuable is the soil archive that contains nearly all soil samples that have been collected between 1962 and 1997. Since about 1990, we have used the soil and plant-tissue archive to estimate changes in the biogeochemistry of soils and the growing forest, and have made a variety of additional measurements to evaluate ecological processes that have affected change in the ecosystem over four decades.

Our objectives in the following four chapters are 2-fold: to describe and interpret how four decades of forest growth have altered soil chemistry; and to use these soil-change data to evaluate difficult-to-measure soil and ecosystem processes involving the cycling of chemical elements including carbon, nitrogen, phosphorus, potassium, calcium, and magnesium. The special feature of the Calhoun experiment is the opportunity it presents to quantitatively evaluate organic carbon turnover, inputs of N_2 fixation and atmospheric nitrogen deposition, changes in organic and mineral fractions of soil phosphorus, soil and ecosystem acidification, and even mineral-weathering releases of calcium, magnesium, and potassium.

For soil organic carbon, the Calhoun study is used to estimate the rate at which 40 years of forest development have accumulated carbon in biomass, forest floor, and mineral soil. We combine these long-term carbon observations with estimates of the inputs of carbon to soil to evaluate turnover of new forest carbon that has been added to soil during the four decades of forest growth. Since the Calhoun forest has grown and developed during a period coincident with a doubling of atmospheric ^{14}C-CO_2 (due to testing of nuclear bombs in the atmosphere), we have a useful isotopic signal for examining accumulation and turnover of new carbon in a soil previously depleted of carbon by its agricultural past.

For soil nitrogen, we describe how four decades of forest growth have transferred about 30% of mineral-soil nitrogen into new biomass and forest floor. Despite the fact that the forest has grown to rely less on mineral-soil nitrogen and more on recycling to meet its nitrogen requirements, the system has grown itself into a state of acute nitrogen deficiency in part due to nitrogen mineralization in the pine forest floor being relatively low. A modest ecosystem accretion of nitrogen (in plants plus soil) is estimated over four decades, and we suggest that the most likely source of the additional nitrogen in the pine ecosystem is from incorporation of atmospheric nitrogen deposition rather than N_2 fixation.

For soil acidity, four decades of decreasing soil pH and increasing exchangeable acidity clearly demonstrate the ecosystem's pattern of acidification. Much of this accumulating acidity is derived from internal ecosystem processes such as biomass accumulation of nutrient cations and carbonic acid leaching. As much as 40% of the accumulation in acidity in the upper 0.6 m of mineral soil may be derived from acids in atmospheric deposition. However, nearly all of the incoming pollutant sulfate is adsorbed by the mineral soil between 0.6 and 1.75 m depth. Acid deposition of sulfate affects a translocation of nutrient cations from surficial soil horizons into the upper B horizons rather than accelerating the loss of nutrient cations from the ecosystem.

For soil potassium, calcium, and magnesium, four-decade depletions (or the lack thereof) in soil exchangeable contents are important observations in themselves. More significant, however, is that observed depletions help us evaluate rates of weathering releases of these nutrients from non-exchangeable pools. At Calhoun, rates of weathering resupply differ greatly among the three cations. For calcium and magnesium, rapid depletion of exchangeable pools in the surficial 0.6 m of soil demonstrates that mineral weathering has limited ability to buffer the soil's bioavailable pools. For potassium, however, removals of soil potassium by uptake plus losses to leaching exceed by nearly 20 times the observed depletions of soil-exchangeable potassium. Bioavailable potassium appears to be amply supplied by releases from non-exchangeable mineral sources, and thus soil and ecosystem cycling of potassium contrasts greatly with that of calcium and magnesium.

Finally for soil phosphorus, four decades of forest development have affected major changes in organic and inorganic soil phosphorus. Agricultural fertilization of cotton prior to 1955 built up soil phosphorus as organo-, calcium, iron, and aluminum phosphates, and these fractions have been slowly drawn upon to meet phosphorus requirements of the growing forest and resupply bioavailable soil phosphorus as well. Bioavailable phosphorus has remained high throughout this period, and thus the Hedley fractionation (Hedley *et al.* 1982a, 1982b) provides a detailed understanding of short- and long-term bioavailability of soil phosphorus. Of all fertilizer nutrients, phosphorus may be most readily targeted to benefit both short- and long-term productivity by recharging both labile and slower-turnover fractions of inorganic and organic phosphorus.

15

Accumulation and rapid turnover of soil carbon in a re-establishing forest

In aggregate, forests of the world are being cut over and converted to non-forest land uses, a transformation that is releasing up to 2×10^{15} g carbon as CO_2 to the atmosphere each year (Sundquist 1993; Oades 1994; Schimel 1995; Schlesinger 1997). At the same time, many secondary forests such as the pine forests of southeastern North America are reaccumulating carbon at rapid rates. Aggrading secondary forests, especially in eastern North America and in Europe, are accumulating carbon in new tree biomass and soils at a rate estimated world-wide to total about 0.5×10^{15} g carbon per year. Although carbon budgets of many types of ecosystems have been studied intensively for several decades, effects of land uses on carbon release and accretion of soils and ecosystems are great approximations (Delcourt and Harris 1980; Melillo et al. 1985; Post and Mann 1990; Schlesinger 1990; Burke et al. 1995; van Lear et al. 1995; Post et al. 1997; Houghton et al. 1998).

Our most precise estimates of land-use effects on soil carbon are based on long-term field experiments that quantify soil changes under controlled regimes of land management (Jenkinson 1991; Leigh and Johnston 1994; Harrison et al. 1995; Smith et al. 1997). In this chapter, we use the Calhoun Experimental Forest to examine four-decade carbon sequestration and turnover in the whole forest ecosystem: in aggrading forest biomass, forest floor, and mineral soil. Soil studies in the vicinity of the Calhoun forest suggest that cropping for cotton and other row crops reduced organic carbon in the upper 0.3 m of the mineral soil by about 40% of the carbon present prior to cultivation (Richter and Markewitz 1996). This estimate was made by comparing carbon content in uncultivated mineral soils that support relatively old, uneven-aged oak

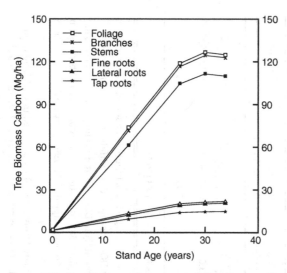

Figure 15.1. Cumulative carbon accumulation in aggrading tree biomass over the first 34 years of forest development at the Calhoun Experimental Forest, SC.

forests with that in previously cultivated mineral soils that currently support pines or row crops (Table 13.2, Figure 13.1).

We also use the Calhoun experiment to argue that well replicated, long-term soil-ecosystem experiments are needed in selected ecosystems throughout the world to examine global carbon dynamics. Although the Calhoun forest has similarities with many forests, forest carbon dynamics may be far different from those observed at Calhoun, due to differences in plant species, soil mineralogy, and climatic regime. To begin to manage the global carbon cycle, soil scientists and ecologists need direct observations of how soils and ecosystems serve as carbon sinks and sources over time scales of decades. Examining carbon accretion in previously disturbed ecosystems is of particular importance to this enterprise (Barber and Van Lear 1984; Johnson and Todd 1998).

REBUILDING ECOSYSTEM AND SOIL CARBON

Following agricultural abandonment in the mid-1950s, the aggrading Calhoun forest has been an effective carbon sink, accumulating nearly 165 Mg ha^{-1} of carbon in 40 years. About 63% of this accretion in carbon is contained in new aboveground tree biomass (foliage, branches, and boles), and about 37% is new organic carbon belowground, in the forest floor, roots, and in the mineral soil. Biomass carbon accumulated most

Table 15.1. *Annual soil-carbon inputs (in 1990s) and estimated 40-year soil-carbon accretions in eight permanent plots of the Calhoun Experimental Forest, SC. Reported are means and in parentheses coefficients of variation (%) among the eight permanent plots (Richter et al. 1999)*

Annual soil-carbon input or soil-carbon accretion	Years of sample	Forest floor	0–15 cm	15–60 cm
			$(\mathrm{Mg\ ha^{-1}\ year^{-1}})$	
Annual canopy litterfall[a]	1991–1992	2.45 (11.9)	—	—
Annual solution DOC input[a]	1992–1994	0.08 (13.1)	0.32 (28.1)	0.19 (34.5)
Annual rhizo-deposition[a]	1994–1995	0.37 (23.0)	0.67 (14.1)	0.20 (26.5)
Annual carbon input	1990s	2.90	0.99	0.39
			$(\mathrm{Mg\ ha^{-1}})$	
Total soil-carbon accretion[b]	1957–1997	37.8 (18.8)	1.45 (50.5)	0

[a]Inputs of carbon in canopy litterfall to the forest floor in monthly collections (Urrego 1993). Inputs of dissolved organic carbon (DOC) were estimated in bi- or tri-weekly collections: into forest floor from DOC in canopy throughfall, to 0 to 0.15 m mineral soil from DOC in water infiltrating from forest floor, and into 0.15 to 0.6 m soil from DOC in water draining into soil at 0.15 m depth (1992–1994). Inputs of carbon from rhizo-deposition were estimated from 50% of mean live fine root (<2 mm) biomass in forest floor, and in mineral soil at 0 to 0.15 m and 0.15 to 0.3 m depths. Fine roots sampled 1994–1995 every three weeks over 18 months.

[b]Soil-carbon accretion over 40 years (from planting in 1957 to 1997) is estimated from forest floor collected in 1997, and from collections of 0 to 0.6 m mineral soil made between 1962 and 1997. Bulk densities for mineral soils are 1.52 Mg m^{-3} in 0 to 0.35 m layers, and 1.44 Mg m^{-3} at 0.35 to 0.6 m.

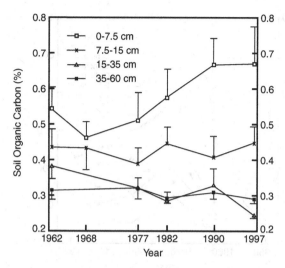

Figure 15.2. Mean concentrations of mineral-soil carbon (1962–1997) in eight permanent plots at the Calhoun Experimental Forest, SC. Plots were planted with loblolly pine seedlings in 1957 following more than 100 years of cultivation for cotton and other row crops. Error bars depict spatial variation (as standard errors) among the eight permanent plots. The randomized complete block ANOVA indicated highly significant increases over the 40 years in soil carbon at 0 to 0.075 m ($P < 0.001$), no significant changes at 0.075 to 0.35 m, and significant decreases at 0.35 to 0.6 m ($P < 0.05$).

rapidly through the first 25 years of forest development (Figure 15.1), and appears to have reached an asymptote in the last decade. The specific explanation of this asymptotic pattern of biomass accumulation is a complex question (Ryan *et al.* 1997), but in the Calhoun forest may be tied to soil-nitrogen depletion (Chapter 16).

Of the new soil carbon in the full profile (soil organic matter in the O horizon plus mineral soil), 37.8 Mg ha^{-1} (CV% = 18.8 among eight permanent plots) is in the forest floor that blankets the formerly cultivated mineral soil, and only about 1.5 Mg ha^{-1} (CV% = 50.5) of carbon has accumulated in humus in the upper layers of mineral soil (Table 15.1). The mineral soil gained carbon significantly in the upper 0 to 0.075 m layer between 1962 and the 1990s ($P<0.001$), but not in soil horizons deeper than 0.075 m (Figure 15.2). In fact, carbon in the lowermost soil layer sampled, at 0.35 to 0.6 m, significantly decreased ($P<0.05$) during the 35-year soil measurement period, 1962–1997, perhaps due to slow oxidation of carbon from crop roots accumulated during the period of farming.

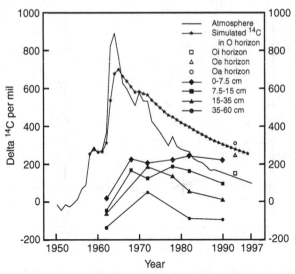

Figure 15.3. Time trends of ^{14}C in atmospheric CO_2 (1950–1997); forest floor of Oi (L), Oe (F), and Oa (H) layers (in 1992); and mineral soil (in 1962, 1968, 1972, 1977, 1982, and 1990) of the eight permanent plots at the Calhoun Experimental Forest, SC. Simulated changes in ^{14}C in O horizons (1957–1996) are estimated from the decomposition model of Jorgensen *et al.* (1980) and estimates of annual litterfall input over the four decades. The Δ^{14}C per mil of the 0.35 to 0.6 m sample in 1962 was estimated by a regression equation of depth vs. Δ^{14}C.

INCORPORATION OF BOMB-PRODUCED ^{14}C-CO_2 INTO SOIL ORGANIC MATTER

Because the Calhoun forest was planted in 1957, the fate of new forest carbon in the Calhoun soil can be examined from the incorporation and disappearance of ^{14}C over the subsequent four decades. New soil carbon derived from the Calhoun forest has a distinctive isotopic label of ^{14}C/^{12}C, because starting in the late 1950s prolific aboveground nuclear-bomb testing nearly doubled the global atmosphere's concentration of ^{14}C in CO_2 (Figure 15.3). Photosynthates in each year's production of new biomass, including litterfall, contain nearly the same ratio of ^{14}C/^{12}C as the CO_2 in the atmosphere. In fact, the atmospheric Δ^{14}C record itself (Figure 15.3) was estimated not by direct measurements of atmospheric CO_2, but by uncorking bottles of wine of annual vintages and determining the Δ^{14}C of the photosynthates contained in each year's wine (Burchuladze *et al.* 1989; Trumbore 1996).

Since a given year's photosynthate (whether in litterfall, rhizo-deposition, or soluble organic compounds in wine) contains $^{14}C/^{12}C$ that closely matches that in the atmosphere, changes in soil radiocarbon (^{14}C) are used to evaluate dynamics of soil organic matter during forest development. The ^{14}C in soil organic matter clearly illustrates how soil organic carbon is not only highly dynamic, but is much more so than might be suggested by the relatively sluggish changes of total soil carbon that have occurred over the four decades of the pine forest's development (Figure 15.2).

SIMULATIONS OF CARBON ACCUMULATIONS IN THE FOREST FLOOR

A decomposition model of Jorgensen et al. (1980) was adapted for simulation of the pattern of ^{14}C in the aggrading forest floor. Simulations suggest that by 1965 the pine-forest $\Delta^{14}C$ approached + 700 per mil in its 8-year-old O horizon (Figure 15.3). The $\Delta^{14}C$ is an index of ^{14}C concentration in organic material, and is simply the deviation in the $^{14}C/^{12}C$ ratio from that of a standard. In the Calhoun pine forest, the peak value of $\Delta^{14}C$ in the forest floor was simulated to be reached only one year after ^{14}C peaked in the atmosphere (Figure 15.3).

By the 1990s, however, simulated $\Delta^{14}C$ in the forest floor had declined to less than + 300 per mil, although its recession greatly lagged the recession in atmospheric ^{14}C-CO_2. The forest floor has incorporated ^{14}C into relatively slow-to-decompose humic compounds over its four decades of accumulation.

Measurements of bomb-produced ^{14}C in the early 1990s indicated that ^{14}C was most concentrated in the basal layers of the O horizon and in the 0 to 0.075 m layer of mineral soil (Figure 15.3). In 1992, the most surficial Oi horizon (the layer of the pine litterfall deposited within three to four years of 1992) had a $\Delta^{14}C$ of + 152.2 per mil, closely similar to that of atmospheric CO_2 during the late 1980s and early 1990s (Figure 15.3). By contrast, the 1992 collections of Oe and Oa horizons (the middle and most basal layers of forest floor) had a $\Delta^{14}C$ of + 247.3 and + 309.8 per mil, respectively (Figure 15.3). These older, most basal layers of forest floor are greatly enriched in ^{14}C derived from plant biomass that was synthesized during the era of greatly elevated atmospheric ^{14}C-CO_2.

Details of a Jorgensen et al. (1980) model that simulates forest floor accumulation and its ^{14}C dynamics are contained in Appendix II.

OBSERVATIONS OF CARBON SEQUESTRATION IN MINERAL
SOIL

Despite the relatively modest change in total organic carbon in the
mineral soil during 40 years of forest development (Figure 15.2, Table
15.1), bomb-produced ^{14}C was rapidly incorporated into the organic
matter *throughout* the upper 0.6 m of mineral soil (Figure 15.3). For
example, by 1968, only four years after the atmosphere's peak in ^{14}C, the
Δ^{14}C of 0 to 0.15 m deep mineral soil had increased to + 200 per mil, up
from − 10.4 per mil in 1962 collections. By 1972, Δ^{14}C of the entire 0 to
0.6 m mineral soil averaged + 125 per mil.

In the several decades following the early 1970s, only mineral soil
at the 0 to 0.075 m depth has maintained an elevated ^{14}C (Figure 15.3).
The soil organic matter in this most surficial, mineral-soil layer is part of
the reaccumulating A horizon that is slowly re-forming under the pine
forest following long-term cultivation. Below 0.075 m, however, short-
term increases in ^{14}C appear to be decreasing in concert with atmospheric decreases in ^{14}C, a pattern that illustrates how forest inputs of
carbon are being decomposed and not sequestered by the mineral soil
(Figure 15.3).

INPUTS AND ACCRETIONS OF SOIL CARBON

To better understand the mechanisms that are behind these soil carbon
dynamics, in the 1990s the forest's several inputs of carbon into the soil
profile were estimated. Inputs included carbon from three main sources:
canopy litterfall, rhizo-deposition (fine root sloughing and exudation),
and hydrologic leaching of dissolved organic carbon (DOC).

Inputs of carbon to the forest floor totaled 2.90 Mg ha^{-1} year^{-1} in
the mid-1990s, 84% of which was from canopy litterfall, with smaller
fractions derived from turnover of fine roots (Vogt *et al.* 1983) and from
DOC leached from the forest canopy (Table 15.1).

The forest's 40-year accretion of carbon in the forest floor, 37.8 Mg
ha^{-1}, is therefore about 13 times the current annual inputs (Table 15.1).
The relatively massive carbon accumulation in the forest floor results
from large inputs of carbon bound in relatively complex, acidic, and
recalcitrant organic compounds that are derived from the coniferous
foliar canopy. The large carbon accumulation in the forest floor is also
attributable to the 40-year exclusion of fire from the pine ecosystem. The
blanket of forest floor is classified as an acidic mor, literally perched
atop the mineral soil's incipient and reaccumulating A horizon, with

relatively little physical incorporation by soil fauna of forest floor material with mineral soil below (Richter and Markewitz 1996).

Inputs of carbon into the 0 to 0.15 m layer of mineral soil totaled about 0.99 Mg ha^{-1} year^{-1} in the mid-1990s, about one-third the carbon inputs to the forest floor (Table 15.1). About 67% of the carbon input to this upper mineral soil was attributed to rhizo-deposition, and the other third was derived from DOC in leachates draining from forest floor layers above (Table 15.1). In contrast with the patterns of carbon sequestration in the forest floor (Table 15.1), the 40-year mineral-soil carbon accumulation, 1.45 Mg ha^{-1}, in the 0 to 0.075 m layer of mineral soil is only 1.5-fold that of the annual inputs in the 1990s. The remarkably short residence time for carbon inputs to these coarse-textured surface soils is particularly notable and has been suggested by others in similar ecosystems (Ruark 1993).

Decomposition in the upper mineral soil is rapid for several reasons. The upper mineral soil is coarse textured (68% sand and 15% clay in the upper 0.35 m of mineral soil), which ensures a high degree of macroporosity and an oxidizing environment. Clay-sized mineral particles that are present are mainly low-activity kaolinite and quartz with little potential to adsorb and physically protect organic carbon from microbial attack. In sum, carbon inputs are lower in mineral soil than in O horizons, and without protection from clay sorption, inputs of DOC and fine roots to mineral soils have not accumulated except at very modest rates.

Decomposition is rapid in the surficial mineral soil, although our so-called "conservative" estimate of rhizo-deposition is highly uncertain and may even be an overestimate. For example, fine roots (<2 mm) may not turn over at the rate suggested in Table 15.1. Samples of <2 mm live fine roots collected in 1998, using the same procedures as used to collect the data for rhizo-deposition reported in Table 15.1, were measured to have a Δ^{14}C of + 145, higher than would be expected if these materials had a 100% turnover per year. In addition, the *net* annual DOC input to the 0 to 0.15 m layer is 0.13 Mg ha^{-1} (0.32 Mg ha^{-1} input from forest floor, 0.19 Mg ha^{-1} output to soil >0.15 m in depth). Thus, the reason for the modest 40-year mineral-soil accumulation of soil organic carbon appears to be not only that decomposition of inputs is rapid, but also that carbon inputs to mineral soils may be relatively lower than might be expected. These dynamics of the carbon cycle are clearly important to resolve in the coming years.

The Calhoun Forest Experiment is currently the only forest ecosystem in which carbon sequestration in the whole system, above- and belowground, has been directly measured in replicated plots over a period of several decades. The system as a whole has sequestered carbon at an enormously high rate: >4.0 Mg ha^{-1} year^{-1} over 40 years. By contrast, mineral soil has been relatively slow to reaccumulate carbon during 40 years of reforestation. Considering that cultivation is estimated to have depleted 40% of the mineral-soil content of carbon, modest changes in mineral-soil carbon during 40 years of vigorous forest regrowth are striking. The abandonment of agricultural disturbance and an influx of forest carbon during a period of vigorous accretion of pine biomass were not sufficient factors to accumulate much organic carbon in mineral soils. Presumably, as this forest matures and makes the transition from pine to hardwoods, additional carbon will be sequestered in mineral soils until a more carbon-rich A horizon is eventually re-formed.

16

Satisfying a forest's four-decade nitrogen demand

Of all plant nutrients, nitrogen is one capable of changing very substantially in soils over time scales of decades. Soil nitrogen can be taken up by plants in potentially large amounts, and microbes can compete effectively for bioavailable nitrogen. Symbiotic microbes can also add nitrogen substantially to soil organic matter via N_2 fixation.

Early successional forests that establish themselves on disturbed landscapes rapidly reorganize an ecosystem's nitrogen cycle (Marks and Bormann 1972; Vitousek and Andariese 1986). As a young pine-forest ecosystem develops, the system meets its immediate nitrogen requirements for growth and also stores nitrogen in slow-to-decompose organic matter that can mineralize nitrogen as the ecosystem ages.

When southern pine ecosystems establish themselves on old agricultural fields or cutover lands, young pine trees initially take up large amounts of soil nitrogen, mainly to produce nitrogen-rich foliar canopies. Secondarily, forests return relatively large amounts of nitrogen back to the soil surface in leaf litter where it can potentially be remineralized for subsequent uptake by plants (Figure 16.1).

Classical studies of southern pine forests (e.g., Switzer and Nelson 1972, Wells and Jorgensen 1975) suggest that mineral soil initially supplies most of the nitrogen needs of old-field pine forests, but that within several decades vegetation is able to meet much of its nitrogen requirement from the mineralization of the forest floor, transfers within the trees' biomass (re-translocation), atmospheric precipitation, and net canopy throughfall (Table 16.1). In no studies to date, however, have we directly observed the cycling of nitrogen during decades-long forest development or how mineral-soil nitrogen responds to forest growth during these highly dynamic decades.

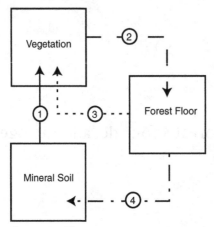

Figure 16.1. Three ecosystem components (two of which are part of the soil, i.e., forest floor and mineral soil) contribute to the retention, cycling, and supply of nutrients in aggrading forest ecosystems growing on disturbed or cutover land. Solid lines indicate processes affecting the forest nutrient cycle early in the life of the secondary forest; dotted lines indicate processes that become increasingly important as the system ages.

This chapter's objectives are to:

- directly estimate accumulations of nitrogen in vegetation biomass plus forest floor and the associated decreases in mineral-soil nitrogen during the first four decades of pine-forest development;
- determine what these soil changes indicate about soil organic matter quality and nitrogen bioavailability during this period;
- estimate the nitrogen accretion in the whole forest ecosystem (i.e., mineral soil, forest floor, plus tree biomass) over four decades and evaluate sources of additional nitrogen that may have accumulated during this period; and
- evaluate the sustainability of nitrogen supply in such a system, and especially its implications for long-term forest management.

NITROGEN ACCUMULATION IN VEGETATION AND FOREST FLOOR

Periodic tree measurements at the Calhoun site indicate that nitrogen accumulation in biomass totaled 366 kg ha^{-1} (CV% = 9.3) over the four decades (Figure 16.2, Table 16.2). The rate of biomass accumulation of nitrogen averaged 17 kg ha^{-1} during the first 15 years, a substantial annual accumulation due in part to the synthesis of the nitrogen-rich

Table 16.1. *Annual nitrogen sources to a 20-year-old loblolly pine ecosystem as estimated by Switzer and Nelson (1972). From this perspective, pine forests were able to sustain their supply of bioavailable nitrogen without productivity becoming limited by nitrogen*

Sources of nitrogen	Nitrogen (kg ha^{-1})	% of total
Litter mineralization[a]	27.9	40
Internal transfer within the biomass	26.9	39
Precipitation	11.3	16
Foliar canopy wash (net throughfall)	3.3	5
Total	69.2	100

[a]Set equal to nitrogen inputs in litterfall based on the assumption of a nitrogen steady state in the O horizon by age 20 years.

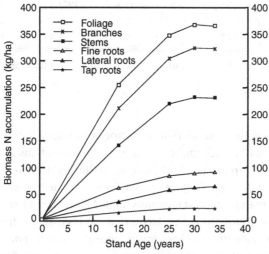

Figure 16.2. Nitrogen accumulation patterns in aggrading biomass over four decades at the Calhoun Experimental Forest, SC (Richter *et al.* 2000b).

foliar canopy. Once the foliar canopy has fully grown, mainly stem wood and bark (with relatively low nitrogen requirement) continue to accumulate biomass and nitrogen (Figure 16.2). Between age 15 and 30, net accumulation of nitrogen in biomass decreased to about 6 kg ha^{-1} year^{-1}, with nitrogen accumulating almost entirely in wood and bark of boles. After age 30, vegetation at Calhoun did not increase in biomass or nitrogen content (Figure 16.2). Tree mortality at this period approximately compensated new biomass increment.

Table 16.2. *Changes in total nitrogen of ecosystem components (vegetation, forest floor, and 0.6 m mineral soil) of eight permanent plots at the Calhoun Forest Experiment, SC (1962–1997). Estimates of ecosystem accretions are also given*

| Plot | 1962 Ecosystem[a] | Mid-1990s | | | Ecosystem accretion |
| | | Biomass | O horizon | Ecosystem[a] | |
			(kg ha^{-1})		
1-8	2022.9	378.6	665.0	2400.7	377.8
2-8	3179.0	300.3	840.3	3106.9	− 72.1
3-8	2418.9	359.3	711.1	2685.4	266.5
4-8	2652.9	408.6	687.7	3299.2	646.3
1-10	3093.9	343.3	673.3	2754.4	− 339.6
2-10	3013.7	384.3	743.2	3415.6	401.9
3-10	2670.0	395.0	847.9	3145.1	475.1
4-10	3081.5	356.4	770.0	2962.9	− 118.6
Mean	2766.6	365.7	742.3	2971.3	204.7
Standard deviation	402.4	34.0	71.9	339.7	341.9
CV%	14.5	9.3	9.7	11.4	167.0

[a]Ecosystem represents the sum of nitrogen in vegetation, forest floor, and 0.6 m mineral soil.

In addition to accumulations of nitrogen in biomass, the forest floor that heavily blankets mineral soil at Calhoun contains an enormous amount of nitrogen (Table 16.2). At age 40 (in 1997), the O horizon contained about 740 kg ha^{-1} of nitrogen (CV% = 9.7). The nitrogen accumulated in biomass plus forest floor thus totaled more than 1100 kg ha^{-1} by the mid-1990s (Table 16.2). One factor that has been absent from the Calhoun forest since 1957 has been fire, a process that could potentially have caused the loss and reduction of carbon and nitrogen from the forest floor (Wells 1971; Richter *et al.* 1982).

DEPLETION OF NITROGEN IN MINERAL SOIL

One of the most striking patterns of soil change observed at the Calhoun forest has been the four-decade depletion in nitrogen in the upper 0.6 m of mineral soil (Figure 16.3), a depletion that parallels nitrogen accumulation in biomass (Figure 16.2) and forest floor (Table 16.2).

Mineral soil was the source for most of the nitrogen accumulated in biomass and forest floor, and over the four decades mineral-soil

Table 16.3. *Contents and depletions of total nitrogen of 0.6 m mineral soil of eight permanent plots at the Calhoun Experimental Forest, SC (1962–1997). Estimated depletion of nitrogen was significantly different from zero with a paired t-test with a probability of <0.0001.*

Plot	1962	1997	Change
		(kg ha⁻¹)	
1-8	1942.9	1357.1	− 585.8
2-8	3099.0	1966.3	− 1132.7
3-8	2338.9	1615.0	− 723.9
4-8	2572.9	2202.9	− 370.0
1-10	3013.9	1737.8	− 1276.1
2-10	2933.7	2288.1	− 645.6
3-10	2590.0	1902.1	− 687.9
4-10	3001.5	1836.5	− 1165.0
Mean	2686.6	1863.2	− 823.4
Standard deviation	402.4	302.9	− 325.0
CV%	15.0	16.3	39.5

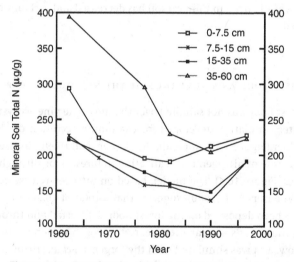

Figure 16.3. Four-decade changes in total nitrogen concentration in mineral soil at the Calhoun Experimental Forest, SC (Richter *et al.* 2000b).

nitrogen decreased by 823 kg ha⁻¹ (CV% = 39.5) (Tables 16.2, 16.3). Total nitrogen in the top 0.6 m of mineral soil amounted to 2687 kg ha⁻¹ in 1962 (CV% = 15.0) and 1863 kg ha⁻¹ in 1997 (CV% = 16.3). The reduction in mineral-soil nitrogen over this time was highly significant

($P < 0.0001$), was observed in all eight permanent plots, and amounted to a reduction to nearly 70% of the initial content in 1962 (Table 16.3).

The observed time series of mineral-soil nitrogen also indicates that almost all of the drawdown in nitrogen occurred during the first 25 years of forest growth, from 1957 to 1982 (Figure 16.3). These initial 25 years of most rapid transfer of nitrogen from mineral soil to biomass and forest floor have been followed by a relatively constant content of soil nitrogen from 1982 to 1997 (Figure 16.3). Averaged throughout the upper 0.6 m of mineral soil, total nitrogen decreased from 301 μg g^{-1} (CV% = 15.0) in 1962 to lows of 192, 178, and 209 μg g^{-1} in 1982, 1990, and 1997, respectively (Figure 16.3).

The depletion of nitrogen from mineral soil has greatly altered the quality of soil organic matter. The C/N ratio, for example, has increased substantially during the four-decade period. In 1962, the C/N ratio averaged 18.8 in the upper 0.15 m of mineral soil, a ratio no doubt reduced by previous agricultural use. Under cotton management (pre-1955), Calhoun soils were enriched by nitrogen fertilization and depleted of carbon by cultivation. During the growth of the Calhoun pine forest, the C/N ratio in the upper 0.15 m of mineral soil has steadily increased to 30.6 in 1990, as nitrogen in mineral soil has decreased and carbon slowly accumulated.

FOREST FLOOR AS A GOVERNOR OF THE NITROGEN CYCLE

Because forest floor was not sampled over the life of the pine ecosystem, the computer simulation developed for examining accretion of ^{14}C in the forest floor (Figure 15.2, Appendix II, Richter et al. 1999) was used to simulate the rate of nitrogen mineralization in forest floor through the four decades (Figure 16.4). The model, based on nitrogen-release coefficients of Jorgensen et al. (1980), suggests that a total of 1024 kg ha^{-1} of nitrogen has been deposited on the forest floor as litterfall and throughfall during the forest's 40-year lifetime. Net nitrogen mineralization is relatively low, and was simulated with the Jorgensen et al. (1980) model to amount to only 12.1 kg ha^{-1} year^{-1} by age 40 (Figure 16.4). The effectiveness with which the pine forest floor has retained formerly bioavailable nitrogen can be gauged by the fact that in 1997, 740 kg ha^{-1} of nitrogen (CV% = 9.7) had accumulated in the forest floor (Table 16.2), about 72% of the simulated 40-year nitrogen inputs. The forest floor is a strong nitrogen sink and hypothetically a governor not only of bioavailable nitrogen but of ecosystem net primary productivity as well.

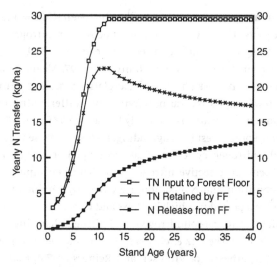

Figure 16.4. Simulation of forest floor nitrogen dynamics based on mineralization coefficients of Jorgensen *et al.* (1980). The model simulated that 697 kg ha^{-1} of nitrogen accumulated in the Calhoun forest by 1997, compared with 740 kg ha^{-1} (CV% = 9.7) that was measured (Richter *et al.* 2000b). No samples of forest floor have been taken and archived from the Calhoun forest before the 1990s. TN, total nitrogen.

ECOSYSTEM NITROGEN ACCRETION

Nitrogen accretion by ecosystems (soil plus vegetation) is highly significant to soil genesis, ecosystem development, and soil management, yet the rate at which ecosystems gain nitrogen has high uncertainty and variability. The scientific literature is particularly intriguing on the topic. The rate at which ecosystems gain nitrogen with and without N$_2$-fixing plants has been studied and debated for many years (e.g., Allison 1955; Stevenson 1959; Jenkinson 1971; Richards 1973; Bormann *et al.* 1993). *Pinus* species have figured prominently in this unsettled literature. Recently, Bormann *et al.* (1993) suggested that several pine ecosystems had increased their nitrogen content by 50 kg ha^{-1} year^{-1}, presumably by associative N$_2$-fixing microbes in the pine rhizospheres.

The Calhoun experiment is well suited for estimating nitrogen accretion by direct comparisons of ecosystem nitrogen in 1962 and 1997, i.e., in mineral soil, forest floor, plus vegetation of eight permanent plots (Table 16.2). In 1962, five years after pine trees were planted at the Calhoun forest, the ecosystem contained a total of about 2767 kg ha^{-1} of nitrogen (CV% = 14.5 among the eight permanent plots). In 1997, the

Calhoun ecosystem contained 2971 kg ha^{-1} of nitrogen (CV% = 11.4) in the same eight plots. The 35-year accretion of ecosystem nitrogen thus totaled about 205 kg ha^{-1}, which a paired t-test estimated to be significantly different from zero with a probability of <0.07. Mean nitrogen gain was estimated to be 5.9 kg ha^{-1} year^{-1}, a relatively modest nitrogen accretion, which we emphasize was not significantly different from zero at $P < 0.05$, although the P-value is relatively low at <0.07 (Table 16.2).

The Calhoun pine forest is an aggrading, nitrogen-deficient ecosystem with a tight nitrogen cycle, one that has helped ensure that the ecosystem has been an effective nitrogen sink. Nitrogen inputs from atmospheric deposition and N_2 fixation are currently retained within the system, during a period in which biomass and forest floor accumulations have depleted mineral-soil nitrogen, and nitrogen leaching losses are remarkably low (Markewitz et al. 1998). This retention of inputs is similar to the "hypothesis" of Vitousek and Reiners (1975), that high biotic demand for nitrogen relative to supply helps ensure effective nitrogen retention by the ecosystem.

Despite the modest rate of nitrogen accretion, the accretion is not likely to be underestimated, and is in fact taken to be an estimate of accumulated inputs from atmospheric nitrogen deposition plus nitrogen fixation. Nitrogen has not been lost by fire, as there has been no burning of the 40-year-old Calhoun forest. Moreover, denitrification and nitrate leaching have likely been of minor significance to the nitrogen cycle of this ecosystem that since 1955 has been unfertilized and increasingly nitrogen deficient. The pattern of soluble nitrogen concentrations in soil water in the 1990s illustrates how little nitrogen has likely been lost to drainage water over the life of the forest. Between 1992 and 1994, concentrations of NH_4 and NO_3 averaged between 25 and 50 μmol L^{-1} in rainfall and throughfall, but once rainwater infiltrated into the soil, soluble NH_4 and NO_3 decreased to only 5 to 10% of concentrations found in water aboveground (Markewitz et al. 1998). Atmospheric nitrogen deposition may be up to 5 to 10 kg ha^{-1} year^{-1}, yet soil leaching through B and C horizons is <1 kg ha^{-1} year^{-1}. Even dissolved organic nitrogen (DON) is not apparently lost from the system in great amounts. Since leaching of dissolved organic carbon (DOC) is <3 kg ha^{-1} year^{-1} (Richter and Markewitz 1996), if a 10/1 C/N ratio is assumed for this DOC, leaching of DON would be <0.3 kg ha^{-1} year^{-1}.

In this ecosystem analysis of nitrogen, the 40-year pine forest has served as a strong nitrogen sink, and the rate of nitrogen accretion approximately equals inputs from atmospheric nitrogen deposition, i.e., 5 to 10 kg ha^{-1} year^{-1}. Since nearly all nitrogen inputs have been re-

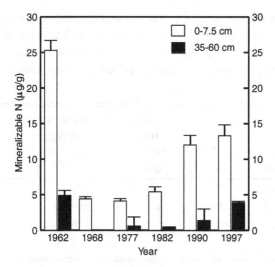

Figure 16.5. Mineralization of nitrogen in aerobic incubations of mineral soils collected in permanent plots at the Calhoun Experimental Forest from 1962 to 1997 (Richter *et al.* 2000b). The incubations were conducted for 30 days at 30 °C in 1999.

tained, nitrogen from hypothetical associative N_2 fixers (Bormann *et al.* 1993) does not appear to contribute much nitrogen to the Calhoun pine ecosystem.

NITROGEN DEFICIENCY IN THE PINE ECOSYSTEM

Low bioavailability of N in the aggrading Calhoun forest is expressed by the substantial decreases in mineral soil total N, the gradually increasing C/N ratio in mineral soil, and the substantial accumulations of N in the forest floor. Low bioavailable N is also expressed by low leaf areas of the foliar canopies, $<2.5\,m^2\,m^{-2}$ when measured in the 1990s (H.L. Allen, unpublished data), and by nitrogen concentrations in foliage of the upper crowns which averaged 1.06% N in the 1990s, low for loblolly pine (Urrego 1993). Aerobic incubations of archived samples of Calhoun soil suggest decreasing mineralizability of mineral-soil N (Figure 16.5), especially during the early years of ecosystem development.

Taken together, these results demonstrate not only that the ecosystem has sequestered relatively large amounts of formerly bioavailable nitrogen but also that it has grown into a state of acute nitrogen deficiency. Nitrogen in mineral soil is reduced to a low and generally

Table 16.4. *Annual sources of nitrogen to the Calhoun forest (1990s), compared with the estimates made by Switzer and Nelson (1972) for old-field loblolly pine forests in Mississippi (Table 16.1). Nitrogen sources may contribute far less nitrogen to the loblolly ecosystem compared with that conceived by Switzer and Nelson (1972)*

Sources of nitrogen	Calhoun nitrogen estimates (kg ha^{-1})	Switzer–Nelson Mississippi estimates (kg ha^{-1})
Litter mineralization	12.1[a]	27.9
Internal transfer within the biomass	24.0[b]	26.9
Total solution inputs to forest floor[c]	8.0	14.6
Total	44.1	69.2

[a]Estimated from litterfall inputs and simulated mineralization rates from data and model of Jorgensen *et al.* (1980). Litterfall was estimated to contain 19.5 kg ha^{-1} of nitrogen in the early 1990s.
[b]Estimated from 43.5 kg ha^{-1} in foliage and 19.5 kg ha^{-1} in litterfall.
[c]Includes atmospheric deposition and net throughfall inputs; measured at Calhoun, a composite estimate in Mississippi.

constant concentration (Figure 16.3), as mineralizable nitrogen in mineral soil was depleted by more than 800 kg ha^{-1}. Trees have increasingly met their nitrogen requirements for annual growth of foliage, wood, and bark not so much from the net transfer of mineral-soil nitrogen to biomass, but from a relatively small pool of bioavailable nitrogen (Table 16.4) that includes 24.0 kg ha^{-1} year^{-1} from re-translocation of nitrogen within trees prior to leaf fall, 12.1 kg ha^{-1} year^{-1} from net nitrogen mineralization in the forest floor (simulated using the model of Jorgensen *et al.* (1980), Figure 16.4), and 5 to 10 kg ha^{-1} year^{-1} from inputs from atmospheric nitrogen deposition (Table 16.4). We hypothesize that vegetation uses most of this nitrogen, which totals 41 to 46 kg ha^{-1} year^{-1}, and that net primary productivity is strongly limited by low bioavailable nitrogen in this ecosystem.

THREE PERSPECTIVES OF SOIL-NITROGEN SUSTAINABILITY

The long-term nitrogen cycle at the Calhoun pine forest leads to three perspectives about sustaining soil-nitrogen supply in these soils that over many years have been managed for both agricultural crops and forest trees.

From the perspective of 20th century harvests of agricultural crops, the nitrogen contained in harvests of cotton and corn were relatively large. In the case of the Calhoun ecosystem, harvests of seed cotton between 1935 and 1955 may have removed about 30 kg ha^{-1} of nitrogen per year from the ecosystem (Mitchell et al. 1996). This rate of nitrogen drain was compensated by nitrogen fertilization, which probably averaged 50 to 100 kg ha^{-1} year^{-1} during this period. Some fertilizer nitrogen not taken up by harvested crops was accumulated in soil organic matter and a fraction was lost as nitrate to groundwater and possibly to denitrification.

From the perspective of wood harvests from the forest ecosystem, conventional logging of the Calhoun pine forest would remove a total of about 140 kg ha^{-1} of nitrogen in stem wood plus bark at age 34 (Urrego 1993). Expressed on an annual basis, this removal from the ecosystem is equivalent to about 4.0 kg ha^{-1} year^{-1} of nitrogen, or < 15% of the annual removal rate in harvests of seed cotton. The removal rate of about 4.0 kg ha^{-1} is less than the atmospheric deposition input of about 5 to 10 kg ha^{-1} year^{-1} (Markewitz et al. 1998) and less than the long-term ecosystem accumulation of nitrogen, so that the 40-year-old ecosystem may not lose nitrogen even if logged. The much greater drain of nitrogen from agricultural ecosystems compared with forestry systems is commonly observed throughout the world.

A caveat for modern forestry systems is that more intensive management increases nutrient drains compared with harvest regimes used in the past. For example, an intensive, complete-tree harvest of aboveground biomass might remove as much as 274 kg ha^{-1} of nitrogen from the Calhoun ecosystem (with 100% biomass utilization), equivalent to a nitrogen drain of 6.9 kg ha^{-1} year^{-1}. Although more intensive forest harvests may well increase nitrogen drain from the ecosystem, the increased drains are likely to remain a small fraction of rates associated with annual agricultural crops.

From the perspective of bioavailable nitrogen in forest soil, however, the relatively low rate of nitrogen drain from periodic forest harvests does not mean that forest growth is not demanding on soil supply of bioavailable nitrogen. Quite the opposite is in fact the case. In the Calhoun forest, the demand for soil nitrogen by the aggrading pine biomass and forest floor is well illustrated by the depletion of total nitrogen in mineral soil (Figure 16.3). Although the nitrogen accumulated in biomass and forest floor is still within the ecosystem, such transfers of

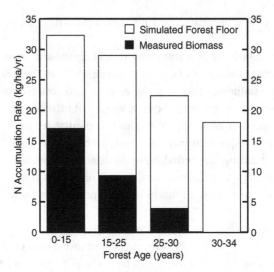

Figure 16.6. Nitrogen accumulation rates in biomass observed by periodic remeasurements of trees in eight permanent plots and in forest floor as simulated with the model of Jorgensen *et al.* (1980).

nitrogen without much return (e.g., without re-mineralization of nitrogen in forest floor) can greatly diminish the bioavailability of nitrogen within the rooting zone (Figure 16.5).

In the 40-year-old Calhoun forest, nitrogen accumulated in tree biomass and forest floor totaled 1108 kg ha^{-1}, equivalent to an accumulation rate of 27.7 kg ha^{-1} year^{-1}. This nitrogen accumulation rate is similar to nitrogen removals in seed-cotton harvests, 30 kg ha^{-1} year^{-1} (Mitchell *et al.* 1996). Without fertilization since the mid-1950s, it is entirely understandable why this rapidly aggrading forest has grown into nitrogen deficiency.

The temporal dynamics of change are likely to be even more significant than is suggested by the mean annual rate of transfer over the four decades. Figure 16.6 illustrates N accumulation in aggrading biomass based on repeated tree measurements and in forest floor based on computer simulations. The transfer of nitrogen from mineral soil to biomass plus forest floor peaks early in the life of the stand, but most especially for biomass. Although the net transfer of nitrogen from mineral soil to forest floor is modeled and not directly observed (Figure 16.6), the low nitrogen mineralization in pine forest floor is a process of considerable importance to pine productivity.

IMPLICATIONS FOR MANAGEMENT

Although the forest's demand for nitrogen outpaces this nutrient's supply in mineral soil, it is significant that the depletion of mineral-soil nitrogen results from a transfer among ecosystem components, and not from a removal of nitrogen from the ecosystem. In contrast to agriculture, silvicultural management has an opportunity to manage a large and aggrading pool of organic matter that during forest development has the capacity to both retain and recycle nutrients. A challenge for intensive pine-forest management is to use fertilization practices in such a way that forest floor and decomposing logging slash are transformed from strong nitrogen sink to regulated nitrogen source, all while minimizing nitrogen loss to drainage water.

Perhaps the most remarkable outcome of nitrogen dynamics in the Calhoun study is that the pine forest was able to take up such a large fraction of its mineral-soil nitrogen. About 30% of the total nitrogen in mineral soils in 1962 (or 823 of 2687 kg ha^{-1}) was transferred from mineral soil and accumulated in biomass and forest floor during the four decades of forest development. Since the rapidly aggrading forest served as a nitrogen sink and losses of nitrogen were relatively small over this period, nitrogen-retention efficiency of the unfertilized Calhoun pine forest is estimated to approach 100% (ecosystem accumulation relative to loss). One of the most important questions for applied soils research concerns how to fertilize ecosystems with nitrogen in such a way as to benefit productivity and maintain high retention efficiency of added nitrogen. Hypothetically, relatively high fertilizer-retention efficiency can be achieved with these pine forests given their high nitrogen uptake, retention of nitrogen in forest floor, extensive root system, and potential for organic nitrogen recycling.

17

Soil re-acidification and circulation of nutrient cations

Farmers and gardeners have long recognized that many crops benefit from soil amendments of lime, marl, ash, and slag (Ruffin 1852; Tisdale *et al.* 1985; Brady and Weil 1996). The ancient Romans and Greeks added calcareous materials to croplands (Tisdale *et al.* 1985). For many centuries, European farmers excavated calcareous substrata that they could spread throughout nearby fields (Johnston *et al.* 1986). In the southeastern USA, Ruffin (1852) articulately promoted calcareous marls for use on crop fields in the Atlantic Coastal Plain.

Today, we know that positive plant responses to additions of lime and other bases are often associated not only with the increase in soil pH, but also with a reduction in metal toxicity, especially that of aluminum, manganese, and iron, and an increase in the bioavailability of phosphorus, calcium, and magnesium (Foy 1984). These edaphic problems associated with excessive soil acidity are well described as an acid-soil infertility complex (Adams 1984). Many agricultural crop plants tend to be susceptible to stress from this complex of problems. By contrast, many tree species that grow and have evolved in acidic soils are relatively tolerant of such conditions.

Soil acidification is now well recognized as a highly significant ecological process and one of the world's major soil-management problems. Soil acidity is also a condition that is so complex that scientific progress in understanding its details was once compared with a merry-go-round (Jenny 1961a). Acidification in natural soil environments has been widely studied in recent years due to concern over the effects of acid deposition on soil acidity (Ulrich *et al.* 1980; Driscoll and Likens 1982; Reuss and Johnson 1986; Tamm and Hallbacken 1986; Binkley *et al.* 1989; Stuanes *et al.* 1992; Markewitz *et al.* 1998).

Prior to forest clearing for agriculture, the Calhoun Ultisols were extremely acidic, and low in exchangeable calcium and magnesium (Table 9.1, Figures 13.5, 13.6). After clearing and conversion to agricultural use, Calhoun soils, like arable acid soils throughout the world, benefited from ash and charcoal inputs from the burned biomass of the primary forest (Table 12.1). Toward the end of the 19th century, lime began to be applied to arable soils of the southeastern USA as a standard agricultural practice (Sheridan 1979), and lime amendments increased in quantity during the early 20th century. Lime applications to the Calhoun soil and to old-field soils throughout the region continue to have residual beneficial effects on base saturation and pH (Figure 13.5) and exchangeable acidity (Figure 13.6), and probably indirectly benefit bioavailability of phosphorus and nitrogen as well.

Calhoun soils have rapidly re-acidified in the decades of reforestation after the additions of agricultural lime were ended in the mid-1950s (Binkley *et al.* 1989; Richter *et al.* 1994; Markewitz *et al.* 1998). The acidification of the Calhoun soil is a process best explained by these Ultisol soils re-adjusting to a formerly acidic state due to the suspension of lime inputs, the regrowth of a forest, and inputs of acid atmospheric deposition. The objective of this chapter is to detail changes in soil chemistry that are directly associated with this recent acidification over several decades.

CALHOUN SOIL ACIDIFICATION AND DEPLETION OF BASE CATIONS

Chemical analyses of soil and trees at the Calhoun forest indicate four important outcomes for soil acidity and nutrient cations during forest development since 1957:

- an acidification of mineral soil that is most rapid in upper layers but that extended throughout the upper 0.6 m of mineral soil;
- a depletion in the upper 0.6 m of soil of about one-half of the exchangeable calcium and magnesium that was present when tree seedlings were planted;
- a soil depletion of calcium and magnesium that is nearly equivalent in magnitude to the calcium and magnesium accumulated in biomass and forest floor plus that leached from the upper 0.6 m layer of mineral soil; and
- remarkably little depletion of exchangeable potassium in the upper 0.6 m of soil, despite the relatively large content of potassium accumulated in biomass and forest floor plus that leached from the upper 0.6 m of soil.

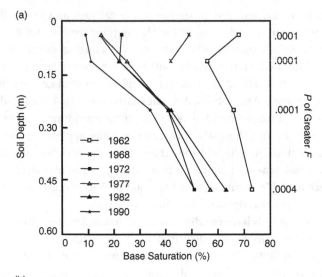

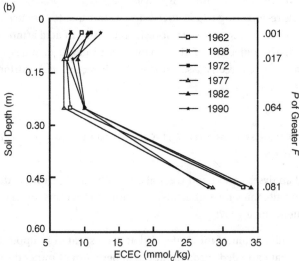

Figure 17.1. Three-decade changes in soil (a) base saturation, (b) ECEC, (c) pH, and (d) exchangeable acidity at four soil depths at the Calhoun Experimental Forest, SC (Richter et al. 1994). Results of an analysis of variance that tested effects of time for each soil depth are indicated by the probability of a greater F.

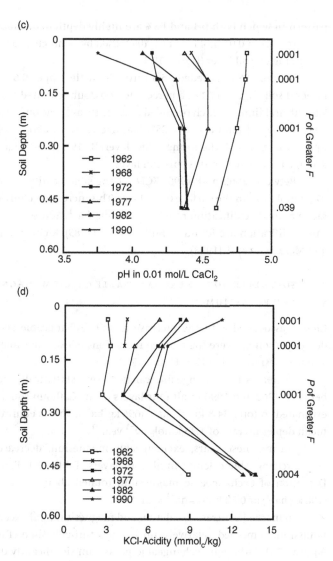

INCREASES IN SOIL ACIDITY

During the growth of the Calhoun forest, decreases in soil pH and increases in exchangeable acidity have been relatively rapid and this acidification appears to have continued through the 1990s (Figure 17.1). Soil pH and base saturation (BS%) have changed from being relatively uniform with soil depth in 1962 (within the upper 0.6 m layer), to a pattern in which both pH and BS% are highly depth dependent (Figure 17.1). Soil pH in 0.01 mol L^{-1} $CaCl_2$ decreased by as much as one pH unit depending on soil layer (Figure 17.1c).

In 1962, BS% exceeded 55% throughout the upper 0.6 m layer of mineral soil (Figure 17.1a), a percentage no doubt elevated by previous agricultural liming which occurred as recently as the growing season of 1954. By 1972 (stand age 15), BS% had decreased to about 20% in the upper 0.15 m of soil, the former plow layer. By 1990 (stand age 34), BS% averaged about 10% in this surficial layer.

Between 1962 and 1990, KCl-exchangeable acidity increased by 37.3 $kmol_c$ ha^{-1} in the surface 0.6 m layer. This increase is equivalent to a mean annual acidification rate of 1.33 $kmol_c$ ha^{-1}. Additional details of this acidification are found in Binkley *et al.* (1989), Richter *et al.* (1994), and Markewitz *et al.* (1998).

SOIL DEPLETION OF EXCHANGEABLE CALCIUM, MAGNESIUM, AND POTASSIUM

Closely associated with the increase in soil exchangeable acidity and decreased soil pH were the marked reductions in exchangeable calcium and magnesium (Table 17.1, Figure 17.2).

Decreases in exchangeable calcium were statistically significant between 1962 and 1990 in all four layers of the Calhoun soil, and were estimated to total 34.8 $kmol_c$ ha^{-1} (696 kg ha^{-1}). This is equivalent to a mean depletion rate of 1.25 $kmol_c$ ha^{-1} $year^{-1}$.

During these years, exchangeable magnesium decreased along with exchangeable calcium in all four layers (Table 17.1, Figure 17.2). Depletion of exchangeable magnesium totaled about 8.9 $kmol_c$ ha^{-1} (108 kg ha^{-1}) or 0.32 $kmol_c$ ha^{-1} $year^{-1}$.

In marked contrast to calcium and magnesium, soil exchangeable potassium remained relatively unchanged through time (Table 17.1, Figure 17.2). Although exchangeable potassium significantly decreased in the 0 to 0.15 m soil layer, these decreases were relatively small. Depletions of exchangeable potassium in these years totaled 0.48 $kmol_c$ ha^{-1} (19 kg ha^{-1}) in the 0.6 m soil layer, or only 0.012 $kmol_c$ ha^{-1} $year^{-1}$.

BIOGEOCHEMICAL PROCESSES THAT DEPLETED EXCHANGEABLE NUTRIENTS

To better evaluate how nutrient demands of the growing forest diminished soil-nutrient supply, soil depletions were compared with processes that removed nutrients from soil (Table 17.2). Nutrient removals from

Table 17.1. *Changes in exchangeable cations and effective cation exchange capacity from 1962 to 1990 in eight permanent plots at the Calhoun Experimental Forest, SC*

Soil layer (m)	Ca	Mg	K	Acidity	ECEC[a]
1962 sample concentrations ($mmol_c\ kg^{-1}$)					
0–0.075	5.00	0.98	0.56	3.1	11.5
0.075–0.15	2.98	0.68	0.42	3.3	7.4
0.15–0.35	3.66	1.05	0.50	2.7	7.9
0.35–0.60	17.50	5.20	1.27	8.9	32.9
1990 sample concentrations ($mmol_c\ kg^{-1}$)					
0–0.075	0.60****	0.23****	0.30***	11.3****	12.6
0.075–0.15	0.47****	0.18****	0.24***	7.4****	8.3*
0.15–0.35	2.27†	0.72*	0.49	6.6****	10.1
0.35–0.60	10.24***	3.15**	1.10	13.4*	27.9
Change in contents 1990 to 1962 ($kmol_c\ ha^{-1}$)					
0–0.075	− 4.9	− 0.84	− 0.28	+ 9.2	
0.075–0.15	− 2.8	− 0.55	− 0.20	+ 4.6	
0.15–0.35	− 3.1	− 0.71	NS	+ 8.7	
0.35–0.60	− 24.0	− 6.78	NS	+ 14.9	
0–0.60	− 34.8	− 8.88	− 0.48	+ 37.3	

*, **, ***, ****Significant at the $< 0.05, 0.01, 0.001$, or 0.0001 probability levels, respectively, for obtaining a higher F for contrasts between 1962 and 1990.
†Significant at the < 0.02 probability level. NS, not significant.
[a]ECEC indicates effective cation exchange capacity (sum of cations), in which NH_4 is included in the most surficial layer, but not included below.

mineral soil (between 1962 and 1990) were estimated from nutrient accumulations in biomass of the aggrading forest, accumulations in forest floor, and losses from the soil's upper 0.6 m due to hydrologic leaching.

Tree biomass accumulated 12.7, 5.5, and 5.7 $kmol_c\ ha^{-1}$ of calcium, magnesium, and potassium, respectively, and the forest floor accumulated 5.7, 1.4, and 1.1 $kmol_c\ ha^{-1}$ (Table 17.2). These accumulations represent relatively rapid removals of cations from mineral soils, especially considering the relatively small inputs from atmospheric deposition and the absence of fertilization during this period (Table 17.3). The rates of accumulation are not dissimilar to other young, rapidly growing *Pinus* stands in the region (Switzer and Nelson 1972;

Table 17.2. *Estimated removals of soil calcium, magnesium, and potassium compared with observed soil depletions (0 to 0.6 m depth) from 1962 to 1990*

| | 28-year change | | |
| | Ca | Mg | K |
Component	$(kmol_c\ ha^{-1})$		
28-year removals			
Vegetation[a]	12.7	5.5	5.7
Forest floor[a]	5.7	1.4	1.1
Net leaching[b]	14.0	7.0	2.8
Total removals	32.4	13.9	9.6
28-year soil depletions[c]	− 34.8	− 8.9	− 0.48
Deficit or surplus	− 2.4	+ 5.0	+ 9.1

[a]Nutrient contents of all vegetation components (foliage, branches, boles, and roots), and corrected for nutrient content (biomass and forest floor) estimated at age five.

[b]Net leaching estimates represent differences between total leaching at 0.6 m depth and atmospheric inputs.

[c]The 1962 and 1990 contents of the 0.6 m layer of soil for exchangeable calcium, magnesium, and potassium were 75.0 and 40.1 $kmol_c\ ha^{-1}$, 21.4 and 12.5 $kmol_c\ ha^{-1}$, and 6.4 and 5.9 $kmol_c\ ha^{-1}$, respectively.

Wells and Jorgensen 1975; Cole and Rapp 1982; Johnson and Lindberg 1992).

Between 1962 and 1990, net soil leaching (gross soil leaching minus atmospheric inputs) from the upper 0.6 m of soil was taken to average 0.5, 0.25, and 0.1 $kmol_c\ ha^{-1}\ year^{-1}$ for calcium, magnesium, and potassium, respectively. These rates were based on three different approaches to estimating long-term soil leaching (Richter *et al.* 1994). Net soil leaching over this period thus totaled 14.0, 7.0, and 2.8 $kmol_c\ ha^{-1}$, respectively (Table 17.3), which for calcium and magnesium approximated accumulations in tree biomass and forest floor (Table 17.2). Based on collections of soil water (1992–1994) and simulation modeling (1957–1990), much of the cation leaching from the upper 0.6 m of soil was associated with sulfate leaching, most of which is attributable to acidic deposition. In the 1990s, sulfate and bicarbonate were 54 and 19% of the anionic charge in soil water at 0 to 0.6 m depth.

Cation deposition from the atmosphere has decreased appreciably in eastern North America over the last several decades (Driscoll *et al.* 1989; Likens *et al.* 1992; Hedin *et al.* 1997), largely following increased air pollutant controls over industrial particulates. Atmospheric inputs of cations are annually small, but may accumulate over decades to become biologically significant to soil cation supply (Låg 1968; Chadwick *et al.* 1999). Atmospheric sources of calcium and magnesium are particularly important to soils with meager weathering releases.

Sulfate leaching removes cations from the upper 0.6 m of soil, but below 0.6 m the soil profile adsorbs nearly all soluble sulfate (Table 17.4). Sulfate concentrations in soil water collected at 0.6 m depth averaged 87.5 $\mu mol_c L^{-1}$ (CV% = 46.7), but only 17.4 and 16.9 $\mu mol_c L^{-1}$ at 1.75 and 6.0 m depths (Table 17.4). Calcium concentrations were closely associated with those of sulfate, decreasing from 70.3 to 8.3 and 9.1 $\mu mol_c L^{-1}$ at 0.6, 1.75, and 6.0 m depths, respectively (Table 17.4, Figure 10.2). Acid deposition is moving nutrient cations lower in the soil but not out of the ecosystem. The hydrous iron oxides in B horizons have great potential to adsorb sulfate (Markewitz *et al.* 1998), and protect the soil from losing calcium and other nutrient cations by leaching. This pattern appears to be typical of other Ultisols (Johnson and Lindberg 1992).

The ecosystem dynamics of the three cations differed greatly in the Calhoun forest. Comparisons of estimated cation removals with observed cation depletions in the upper 0.6 m of mineral soils (Table 17.2) indicated that depletions of exchangeable calcium were approximately equal to estimated removals. In other words, observed soil depletions in the upper 0.6 m of soil were comparable to calcium contents that were accumulated in tree biomass and forest floor, plus that lost from the upper 0.6 m by net leaching.

During the 1962–1990 period, observed magnesium depletions in this 0.6 m of soil were exceeded by estimated magnesium removals by 1.5-fold (Table 17.2). Since depletions of exchangeable calcium and magnesium were on the same order of magnitude as soil removal, removals of calcium and magnesium from the upper 0.6 m soil layer appeared to outpace mineral-weathering releases of these cations plus biological recycling.

Dynamics of exchangeable potassium, however, contrasted markedly with those of calcium and magnesium. Over the three decades, soil potassium removals exceeded by nearly 20-fold the observed depletions in the upper 0.6 m (Table 17.2). Non-exchangeable potassium was clearly able to supply plant-available potassium to the aggrading forest.

Both chemical and mineralogical analyses indicated that the Calhoun soil profiles contain minerals potentially able to release potassium to bioavailable forms (Richter *et al.* 1994; Markewitz 1996; Markewitz and Richter 2000). X-ray diffraction (XRD) indicated the presence of

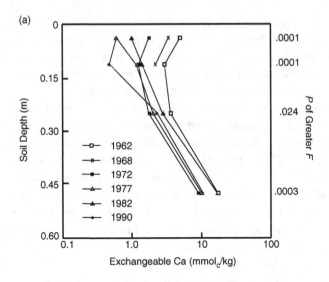

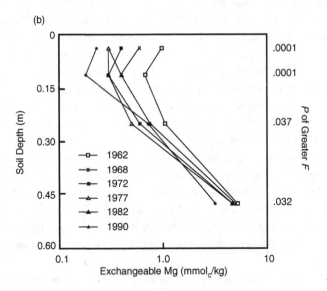

Figure 17.2. Three-decade changes in soil exchangeable (a) calcium, (b) magnesium, and (c) potassium at four soil depths at the Calhoun Experimental Forest, SC (Richter *et al*. 1994). Results of an analysis of variance that tested effects of time for each soil depth are indicated by the probability of a greater *F*.

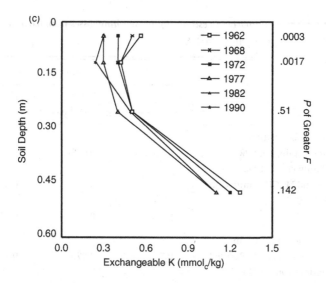

hydroxy-interlayered vermiculite (HIV) in the upper 4 m of the soil profile, especially in clay-sized fractions (<2 μm). Heat treatments of potassium-saturated clay samples only partially collapsed XRD peaks from 1.4 to 1.0 nm, a behavior that indicates that the Calhoun vermiculites contain substantial interlayers of Al hydroxides that potentially contain sites for K (Harris *et al.* 1988; Barnhisel and Bertsch 1989; Douglas 1989; Fanning *et al.* 1989). The X-ray diffraction also suggests the presence of potassium-containing micaceous materials (with 1 nm XRD peaks) in <2 μm, 2 to 45 μm, and >45 μm soil materials from various layers of the soil profile (Markewitz and Richter 2000). Mineralogical research emphasizes the importance of silt-size fractions in soil potassium supply (Haagsma and Miller 1963; Alexiades *et al.* 1973; Comerford *et al.* 1990; Portella 1993). Finally, large micaceous-looking flakes (>45 μm) were physically removed from soil samples taken from the lower soil depths (>4 m) and these flakes were observed to have large 0.33 and 1 nm XRD peaks before and after heating to 550 °C, suggesting the presence of potassium-containing muscovite mica (Figure 17.3).

The presence of potassium-containing minerals was also supported by soil extractions with 1 M HCl (Sparks 1987). About 30-fold more potassium was recovered by these acid treatments than was found in exchangeable pools, and this pattern was particularly prominent within B and C horizons below 2 m soil depth (Figure 17.4). Given the multiple sources of mineral potassium found throughout the Calhoun soil profile, it is likely that some mineral potassium can be released to soil exchange sites, plant roots, and microbes (Markewitz and Richter 2000).

Table 17.3. *Three approaches to estimating net leaching losses from 0.6 m soil at the Calhoun Experimental Forest, SC*

	Elemental fluxes		
Approach	Ca	Mg $(kmol_c\ ha^{-1}\ year^{-1})$	K
Atmospheric deposition			
Solution sampling and PROSPER simulation (1962–1990)	0.072	0.028	0.012
PROSPER and MAGIC simulation (1962–1990)	0.102	0.033	0.012
Literature[a]	0.243	0.056	0.058
Gross soil leaching at 0.6 m			
Solution sampling and PROSPER simulation (1962–1990)	0.590	0.279	0.100
PROSPER and MAGIC simulation (1962–1990)	0.693	0.302	0.102
Literature[a]	0.657	0.227	0.165
Net soil leaching at 0.6 m			
Solution sampling and PROSPER simulation (1962–1990)	0.518	0.251	0.088
PROSPER and MAGIC simulation (1962–1990)	0.592	0.270	0.090
Literature[a]	0.414	0.171	0.106

[a]Johnson *et al.* 1988; Johnson and Lindberg 1992.

PLANT PRODUCTIVITY AND BIOAVAILABILITY OF NUTRIENT CATIONS

In the South Carolina soils, no lime or fertilizer nutrients have been added since at least 1954, and these Ultisols are thus readjusting to a more inherently acidic, pre-cultivation soil condition (McCracken *et al.* 1989). The rapid decrease in soil calcium and magnesium supplies suggests that multiple rotations of productive forests are not sustainable on this soil without fertilizer additions of these nutrients (Richter *et al.* 1994). This is a similar lesson to that learned by 19th and early 20th

Table 17.4. *Annual volume-weighted concentrations of major chemical constituents in precipitation, throughfall, and soil water at the Calhoun Experimental Forest, SC (1992–1994)*

Solution type	H	Ca	Mg	K	Na	NH$_4$	Alk[a]	Cl	NO$_3$	SO$_4$	ΣC[a]	ΣA[a]	DOC[a]
							(μmol$_c$ L^{-1})						(mg L^{-1})
Wet-only precipitation	34.9	7.2	2.4	1.0	8.8	9.0	0.0	18.0	15.5	32.5	63.3	66.0	1.8
Throughfall	39.5	22.1	11.4	21.7	13.0	9.8	0.0[b]	24.7	22.5	54.5	117.5	101.7	7.7
Forest floor leachate	27.4	67.3	47.1	71.9	20.6	34.7	0.0[b]	33.6	10.1	61.6	269.0	105.3	31.4
0.15 m soil water	11.6	66.4	43.5	38.8	27.8	8.4	2.4[b]	33.6	3.1	68.6	196.5	107.7	19.2
0.6 m soil water	1.1	70.3	32.7	12.7	35.6	1.1	32.9	41.8	2.5	87.5	153.5	164.7	1.4
1.75 m soil water	1.2	8.3	9.0	9.1	73.6	0.7	28.3	59.8	3.6	17.4	101.9	109.1	1.0
6.0 m soil water	0.9	9.1	9.8	18.8	75.0	0.3	33.4	64.6	6.2	16.9	113.9	121.1	0.8

[a] Alk is alkalinity, ΣC and ΣA are sums of measured cations and anions, and DOC is dissolved organic carbon. The differences between measured cations and anions in throughfall, forest floor leachate, and 0.15 m soil water are attributed to organic anions, 5.8, 163.7, and 88.8 μmol$_c$ L^{-1}, respectively.

[b] Median is presented for highly skewed distributions where 88, 82, and 48% of all values are zero for throughfall, forest floor, and 0.15 m soil water collections, respectively.

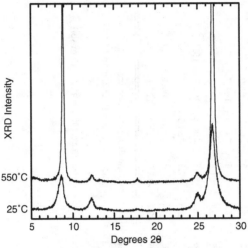

Figure 17.3. X-ray diffractogram of apparent mica flakes washed from 4 to 8 m soil layers at the Calhoun Experimental Forest, SC (site P-1). The 8.8 degree peak that is enhanced by 550 °C is an indication of muscovite mica (Markewitz and Richter 2000).

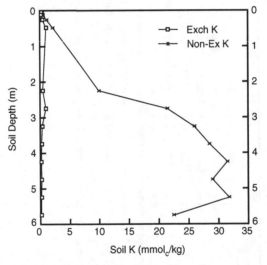

Figure 17.4. Pattern of exchangeable and non-exchangeable potassium (extractable in 1 mol L^{-1} HCl) in the 0 to 6 m layer of soil at the Calhoun Experimental Forest, SC (Richter et al. 1994).

century farmers on southeastern USA sites supported by Ultisols. The perspective with soil potassium is more sanguine. Micaceous and HIV minerals appear to be able to resupply potassium in bioavailable form at

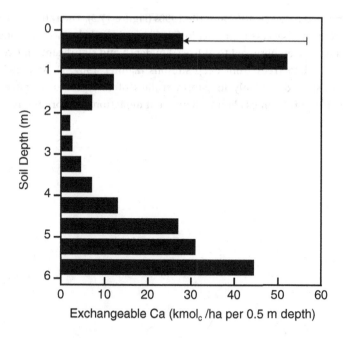

Figure 17.5. Exchangeable calcium in the 0 to 6 m soil profile at the Calhoun Experimental Forest, SC in 1990 (Richter *et al.* 1994). The 1962–1990 depletion of exchangeable calcium in the 0 to 0.6 m layer is illustrated with the arrow.

rates that can keep pace with nutrient demands of forest growth (Markewitz and Richter 2000).

The rate at which nutrient-cation depletion adversely affects forest productivity is not at all clear. Although soil exchangeable calcium is relatively meager, soil calcium supply is by no means exhausted. Bio-available calcium is found mainly in three parts of the soil (Figure 17.5): the O horizon, where the biotic accumulation of calcium is available for recycling and tree uptake; the upper B horizon, likely due to residual effects from agricultural liming (McCracken *et al.* 1989; Richter *et al.* 1994); and the lower C horizons, apparently a weathering front above granitic-gneiss bedrock that underlies the Calhoun soil at an uncertain depth (Calvert *et al.* 1980; Stolt *et al.* 1992; Richter and Markewitz 1995b).

Unlike the case for potassium, there are no indications of large pools of non-exchangeable calcium and magnesium minerals in the profile (Markewitz and Richter 2000), and multiple tree rotations would presumably be able to deplete exchangeable calcium and magnesium from throughout the A, E, and upper B horizons following the pattern established between 1962 and the 1990s (Figure 17.5). The rate at which tree roots may access nutrients in C horizons is little known. Although roots have been observed to at least 4 m depth and significant moisture is supplied to trees from deep subsoils (Zahner 1967; D. Markewitz, unpublished data), only long-term studies of forested soils and ecosystems can inform us about soil-nutrient depletions and productivity.

18

Changes in soil-phosphorus fractions in a re-establishing forest

In the 1850s, E.W. Hilgard (1860) rode on horseback across the state of Mississippi, evaluating the ecology and geology of the state's natural regions and sampling soil for chemical and physical analyses. Hilgard subsequently evaluated soil-chemical concentrations by extracting soil samples with mineral acids. Considering that we still extract soil phosphorus with mineral acids, we have not moved very far beyond Hilgard's methods of soil analysis, despite the passage of nearly 150 years.

Because most soil phosphorus is not available for immediate use by plants or microbes, there have been many attempts to quantify relationships between bioavailable and less-available phosphorus (Thomas and Peaslee 1973; Buol 1995). Nearly all soil and ecological studies that evaluate "available soil phosphorus" use extractants that are correlated with yields of agricultural crops grown in the same year as that of the soil analysis (Bray and Kurtz 1945; Mehlich 1978; Olsen *et al.* 1954). Such correlations are motivated by the prediction of phosphorus-fertilizer recommendations, and are well suited to the dose–response approach to soil management (Thomas and Peaslee 1973).

Results from one such phosphorus-availability test are illustrated for the soil samples in the Calhoun archive (Figure 18.1). The extraction, known as Mehlich III, is composed of EDTA and acetic, hydrofluoric, and nitric acids, and was originally designed for acidic highly weathered soils such as those at the Calhoun forest (Mehlich 1978). The data indicate that surficial soil layers (0 to 0.15 m) have especially high concentrations of available phosphorus (extractable in Mehlich III solution) and that subsoils (0.35 to 0.6 m) have extractable phosphorus that is barely detectable (Figure 18.1). It appears that fertilization, last applied to cotton at the site in 1954, mainly enhanced soil phosphorus in the old

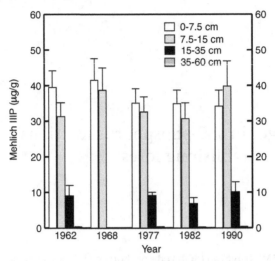

Figure 18.1. Pattern of change in Mehlich III-extractable phosphorus at the Calhoun Experimental Forest, SC. Means and standard deviations are reported ($n = 8$ permanent plots). No significant changes were detectable in these patterns over 28 years. No samples of soil from greater than 15 cm depth were collected in 1968.

plow layer, of about 0.15 m depth. Soil contents of the Mehlich III phosphorus are also relatively constant during the 28 years of forest growth (Figure 18.1), and total 109 and 116 kg ha^{-1} in 1962 and 1990, respectively. In comparison, trees and forest floor had accumulated about 82.5 kg ha^{-1} of phosphorus by the early 1990s, nearly all of which had been taken up from the mineral soil.

The objective of this chapter is to evaluate short- and long-term bioavailability of soil phosphorus over three decades of forest development. The Calhoun experiment is well suited for using the fractionation procedure of Hedley et al. (1982a, 1982b) to examine how soil organic and inorganic compounds of phosphorus respond to long-term plant uptake of bioavailable phosphorus. In long-term agricultural studies, phosphorus fractions have been indicated to be sinks or sources for bioavailable phosphorus, as phosphorus fractions increase or decrease over time (Schmidt et al. 1996). Applied to the Calhoun ecosystem, a system unfertilized since prior to 1955, the uptake and accumulation of phosphorus in tree biomass and forest floor between 1957 and 1990 has hypothetically resulted in the drawdown of relatively stable organic and inorganic fractions of soil phosphorus. The relatively high and consistent phosphorus concentrations of Mehlich III extractions suggest that major transformations have occurred in less bioavailable fractions of soil phosphorus.

The development of soil phosphorus fractionation has included kinetic studies with associated phosphate-sorption isotherms (Fox and Kamprath 1970), models to predict changes in extractable phosphorus over time and management (Cox et al. 1981), and studies that demonstrate long-term effects of phosphorus fertilization on soil-extractable phosphorus and crop yield (Barber 1979; Cope 1981; McCollum 1991; Schmidt et al. 1996). The sequential extraction procedure of Hedley et al. (1982a, 1982b) is based on a wealth of research on soil-phosphorus chemistry (e.g., Chang and Jackson 1957; Shelton and Coleman 1968; Dalal 1977; Novais and Kamprath 1978), and the method has been particularly useful in estimating soil-phosphorus fractions that contribute to short- and long-term bioavailability in a wide range of soils (Tiessen and Moir 1993; Cross and Schlesinger 1994). Recent application of the method to long-term studies has provided new information about the dynamics of soil-phosphorus chemistry over time scales of decades (Goh and Condron 1989; Tiessen et al. 1992; Beck and Sánchez 1994; Schmidt et al. 1996, 1997).

In principle, the Hedley method (Hedley et al. 1982a, 1982b) removes progressively less-available phosphorus with each subsequent soil extraction, and it provides data on both organic and inorganic soil phosphorus (P_o and P_i, respectively) as well. Hedley extractants characterize soil-phosphorus compounds such as labile and exchangeable P_i; desorbable P_i at pH 8.5; iron- and aluminum-associated P_i; calcium-associated P_i; and two fractions of mineralizable P_o, one more readily mineralizable than the other.

The Hedley fractionation procedure used with Calhoun soils follows closely that detailed by Tiessen and Moir (1993). Samples of soil were sequentially subjected to a series of 16 h extractions. The extracts in sequential order were: 0.5 M $NaHCO_3$ at pH 8.5; 0.1 M NaOH; 1 M HCl; concentrated HCl at 80 °C; and digestion with concentrated H_2SO_4 at 360 °C. Total and inorganic phosphorus (with organic phosphorus estimated by difference) were determined in $NaHCO_3$, NaOH, and concentrated HCl extractions; total phosphorus was determined in the 1 M HCl and H_2SO_4 extractions. Inorganic phosphate in $NaHCO_3$, NaOH, and concentrated HCl extractions was determined colorimetrically (Murphy

Table 18.1. *Changes in soil-phosphorus fractions (Hedley* et al. *1982a, 1982b)*
from 1962 to 1990 in eight permanent plots at the Calhoun Experimental Forest,
SC. Means and standard deviations are reported; the generalized analysis of
variance table is given in Appendix III. An important modification to the Hedley
protocol was that resin extracted phosphorus separately, not sequentially prior
to NaHCO$_3$. We report here inorganic phosphorus extracted by NaHCO$_3$
corrected for concentrations recovered by the resin

Soil-P fraction and depth (m)	1962 Mean	SD	1990 Mean	SD	Probability of > F
		$(\mu g\ g^{-1})$			
Resin[a]: Labile P$_i$					
0–0.075	12.6	2.09	12.9	3.16	0.817
0.075–0.15	10.3	3.45	11.1	3.32	0.079
0.15–0.35	3.97	0.65	3.86	1.53	0.755
0.35–0.60	1.85	0.34	0.82	0.25	0.003*
Inorganic 0.5 M NaHCO$_3$[a]: Labile P$_i$					
0–0.075	9.23	3.66	10.51	4.75	0.242
0.075–0.15	6.76	6.82	15.04	10.33	0.059
0.15–0.35	1.05	1.34	4.84	3.40	0.007*
0.35–0.60	1.59	1.12	2.65	1.00	0.018*
Organic 0.5 M NaHCO$_3$: Readily mineralizable P$_o$					
0–0.075	8.30	0.40	6.69	2.47	0.236
0.075–0.15	7.39	2.39	5.89	2.64	0.175
0.15–0.35	3.43	1.01	3.25	1.43	0.715
0.35–0.60	1.82	0.89	1.49	0.98	0.198
Inorganic 0.1 M NaOH: Fe- and Al-associated P$_i$					
0–0.075	39.0	7.54	26.0	8.12	0.014*
0.075–0.15	35.6	6.19	37.1	7.62	0.271
0.15–0.35	20.6	3.79	24.4	2.95	0.0001*
0.35–0.60	46.8	12.67	41.9	12.02	0.154
Organic 0.1 M NaOH: Mineralizable P$_o$					
0–0.075	25.9	8.35	16.5	6.83	0.025*
0.075–0.15	29.1	6.07	18.6	6.04	0.015*
0.15–0.35	18.6	3.42	16.2	2.99	0.322
0.35–0.60	13.5	7.17	13.6	7.60	0.706
1 M HCl: Ca-associated P$_i$					
0–0.075	7.49	4.29	2.62	1.01	0.011*
0.075–0.15	4.95	2.15	3.14	1.08	0.006*
0.15–0.35	1.23	0.49	1.07	0.59	0.295
0.35–0.60	0.48	0.22	0.54	0.26	0.159

Table 18.1. (*cont.*)

Soil-P fraction and depth (m)	1962		1990		Probability of $> F$
	Mean	SD	Mean	SD	
		$(\mu g\ g^{-1})$			
Inorganic concentrated HCl: Ca-associated P_i					
0–0.075	11.8	5.00	8.7	2.88	0.088
0.075–0.15	14.5	5.05	10.1	2.89	0.105
0.15–0.35	16.8	5.28	15.2	5.81	0.380
0.35–0.60	57.8	18.06	46.3	15.86	0.049*
Organic concentrated HCl: P_o					
0–0.075	1.23	0.80	1.25	1.00	0.418
0.075–0.15	1.24	0.75	1.36	1.23	0.854
0.15–0.35	1.20	0.79	1.20	0.63	0.989
0.35–0.60	1.84	0.93	1.96	2.74	0.382
Residual: Recalcitrant P					
0–0.075	19.7	9.61	22.3	14.49	0.233
0.075–0.15	20.0	6.54	25.2	15.62	0.455
0.15–0.35	30.9	14.79	36.7	24.12	0.329
0.35–0.60	74.6	24.00	82.2	25.11	0.531
Total recovery					
0–0.075	135.3	31.33	107.3	29.35	0.034*
0.075–0.15	129.8	15.16	128.9	19.10	0.883
0.15–0.35	97.6	18.16	106.7	26.47	0.171
0.35–0.60	200.2	52.03	189.7	53.73	0.619
Independent total					
0–0.075	143.9	29.97	118.0	32.83	0.096
0.075–0.15	152.0	25.07	131.7	28.35	0.037*
0.15–0.35	123.7	28.03	120.0	26.30	0.654
0.35–0.60	257.7	68.33	228.2	63.56	0.004*

*Significant difference with probability of < 0.05.

[a]Resin and $NaHCO_3$ extractions were performed separately, not in sequence, and results of resin phosphorus have been subtracted from recoveries of inorganic phosphorus in $NaHCO_3$ extractions. For example, $NaHCO_3$ extractions recovered 21.8 and 23.4 $\mu g\ g^{-1}$ of inorganic phosphorus in the 1962 and 1990 samples of soil from 0–0.075 m depth, respectively.

and Riley 1962). Total phosphorus in extracts was determined after digestion with ammonium persulfate and H_2SO_4 (Environmental Protection Agency 1971), and organic phosphorus was estimated by difference. Separate samples were analyzed for total phosphorus by means of a nitric–perchloric acid digestion to compare with recovery of phosphorus by the summation of all component fractions.

One difference between our fractionation of Calhoun soils and that described by Tiessen and Moir (1993) is that resin extraction was performed separately rather than sequentially in the fractionation. The labile resin-extractable inorganic phosphorus is assumed to be entirely contained in the inorganic phosphorus recovered in the $NaHCO_3$ extract.

DYNAMICS OF SOIL-PHOSPHORUS FRACTIONS IN THE CALHOUN FOREST

Compared with other macronutrients, phosphorus is an element of notably low solubility. During the development of the pine forest, leaching and erosion have not removed much phosphorus from the ecosystem. Since 1957, nearly all of the agricultural phosphorus fertilizer has remained in the forest ecosystem, where much of it has circulated between soils and plants.

Between 1962 and 1990, no significant changes were found in labile soil phosphorus extractable by resin strips or in 0.5 M $NaHCO_3$ in the 0 to 0.15 m soil layer (Table 18.1). At depths of 0.15 to 0.6 m, phosphorus extractable in $NaHCO_3$ significantly increased (albeit by small amounts). Although resin-extractable phosphorus significantly decreased by a relatively small concentration at 0.35 to 0.6 m depths (from 1.85 to 0.82 $\mu g\ g^{-1}$), the subsoil layers acted as a phosphorus sink for $NaHCO_3$-extractable phosphorus, possibly caused by decreased soil pH and increased phosphate adsorption during this period (Figure 17.1). Hypothetically, relatively recalcitrant fractions of soil phosphorus, which had been recharged by prior agricultural fertilization, were slowly being drawn down not only in meeting the demands of the aggrading forest but also in maintaining relatively high concentrations of labile soil phosphorus.

Soil organic phosphorus extractable in 0.1 M NaOH was one fraction drawn upon to meet uptake demands of the growing forest (Table 18.1). Between 1962 and 1990, significant decreases were observed in organic phosphorus extractable by NaOH in the two uppermost soil layers sampled (0 to 0.075 and 0.075 to 0.15 m). These decreases were

Table 18.2. *Estimated removals of soil phosphorus between 1962 and 1990 compared with statistically significant depletions or surpluses in Hedley fractions of soil phosphorus (0 to 0.6 m depth)*

Phosphorus flux or component		28-year change in P (kg ha^{-1})
Mineral-soil removals or additions		
Vegetation[a]		− 39.4
Forest floor[a]		− 45.9
Net leaching[b]		+ 2.8
Total		− 82.5
Observed soil change		
Resin P	Labile, exchangeable	− 3.7
NaHCO$_3$ P$_i$	Labile, adsorbed	+ 11.5
NaOH P$_i$	Fe and Al associated	− 3.3
1 M HCl Pi	Ca associated	− 7.6
Conc. HCl Pi	Ca associated	− 41.4
NaHCO$_3$ P$_o$	Readily mineralizable	0
NaOH P$_o$	Stable but mineralizable	− 22.7
Residual P	Organic and mineral recalcitrant	0
Total P	Sum of significant changes	− 67.2

[a]Nutrient contents of all vegetation components: foliage, branches, boles, and roots; corrected for relatively small phosphorus contents in vegetation (5.6 kg ha^{-1}) and forest floor (1.1 kg ha^{-1}) at age five in 1962 (Switzer and Nelson 1972).
[b]Net leaching estimates represent differences between atmospheric inputs and gross leaching at 0.6 m depth, and estimated conservatively to be + 0.1 kg ha^{-1} per year accumulation (Johnson and Lindberg 1992).

equivalent to about 28% of the phosphorus removed from the soil by the aggrading forest between 1962 and 1990 (Table 18.2). Inorganic phosphorus associated with iron and aluminum that is extracted by high-pH NaOH also contributed significant though smaller amounts of phosphorus (Table 18.2). Hedley fractionations of agricultural soils of other long-term studies support patterns observed at Calhoun: that organic and inorganic phosphorus extracted by NaOH is an important long-term source for more labile fractions that are extractable by resin and NaHCO$_3$ (Schmidt et al. 1996, 1997). Moreover, phosphorus contained in soil organic matter that is solubilized by 0.1 M NaOH is considered to have a slower turnover time than organic phosphorus extractable by NaHCO$_3$ (Tiessen and Moir 1993; Cross and Schlesinger 1994).

Calcium-associated phosphorus that is extractable in strong acids was the second major fraction drawn upon to meet uptake demands of the growing forest (Table 18.2). Decreases in 1 M HCl extractable phosphorus were significant in surface layers, and decreases in concentrated HCl fractions were much more substantial at depth (Table 18.1). Concentrated-acid extractions of soil phosphorus have declined in association with the four-decade acidification of the Calhoun soil (Figure 17.1), as insoluble calcium phosphates that were accumulated during the period of phosphorus fertilization and liming (pre-1954) were slowly brought into the biological cycle. Decreases in calcium-associated phosphorus amounted to nearly 62% of the phosphorus removed from soil by the aggrading forest (Table 18.2).

Several studies which have used Hedley fractionations to examine long-term changes in soils indicate that phosphorus removed in crop harvests is associated with decreased resin and $NaHCO_3$-extractable phosphorus. Phosphorus has, however, rarely been completely depleted from these fractions, even in crop systems that have especially large crop uptake (Hedley *et al.* 1982a; Tiessen *et al.* 1983; Schmidt *et al.* 1996). In the Calhoun forest ecosystem, phosphorus uptake by trees and accretion in the forest floor (Table 18.2) totaled about 2.5 kg ha^{-1} year^{-1} (mean annual increment over the 28 years of 1962 to 1990), much less than crop removals of perhaps 5 to 30 kg ha^{-1} year^{-1} for cotton or corn (Mitchell *et al.* 1996; Schmidt *et al.* 1997). As a result, organic and inorganic phosphorus fractions appear to have readily resupplied more labile fractions of phosphorus that in turn have been taken up by roots and microbes.

At the Calhoun forest, the relatively large reductions in organic phosphorus in surficial horizons and in calcium-associated phosphorus at depth are more permanent than changes in labile fractions. Given the low solubility of, and appreciable biotic demands for, soil phosphorus, inputs of fertilizer phosphorus appear readily manageable with high efficiency of retention in forest systems. In the future, phosphorus fertilization of managed forests (and crops as well) should be targeted both at short-term productivity gains and at recharging slower-turnover inorganic and organic fractions. The Hedley fractionation offers the potential to evaluate slow-turnover fractions that over time release phosphorus to more labile, bioavailable fractions and thus to plant-root and microbial uptake.

One significant drawback to the Hedley analytical procedure is the complexity and difficulty of the methods (Tiessen and Moir 1993). An abbreviated, more operational Hedley method seems entirely possible in which three sequential extractions (0.5 M $NaHCO_3$, 0.1 M NaOH, and

12 M HCl) will hypothetically provide most of the information provided by complete fractionation but with substantially less time and effort. Such a three-stage soil analysis could provide a fraction of labile and bioavailable phosphorus, as well as phosphorus bound to iron/aluminum oxides and calcium, plus two organic fractions of different mineralizability. The Hedley fractionation has proven useful to research. Streamlining the Hedley analytical procedures would help greatly in bringing the method into more operational use.

Part V

Soil change and the future

Pertaining to long-term ecological field studies:

"Such a method requires concerted action such as is unknown at present, but there can be little question that continuous investigations of this nature will soon be organized by great botanical institutions."

F.E. Clements (1916)

19

The case for long-term soil-ecosystem experiments

Our understanding of temporal soil change lags far behind that of other components of terrestrial ecosystems. The notably poor understanding of soil change, over time scales of decades, limits ecosystem analysis and our ability to improve land management. With this book, changes in advanced weathering-stage soils have been described and evaluated over time scales of millennia, centuries, and decades.

Over time scales of millennia, many upland soils such as those at the Calhoun forest have limited capability to accumulate organic carbon and nitrogen due to coarse texture and low-activity clay minerals. Over these time scales, low inputs of nitrogen are assumed to be balanced by similarly low removals. Natural processes of soil formation acidify advanced weathering-stage soils (Richter and Markewitz 1995b). These processes deplete primary minerals in soils on geomorphically stable landforms of humid environments (Tables 1.1, 9.1, Figure 3.2). Weathering release of many nutrients is diminished over time, including that of phosphorus, calcium, magnesium, and potassium. Although nutrient removals may outpace inputs for many nutrient elements, bioavailability of some nutrients such as potassium and phosphorus may be buffered by formation of secondary minerals that resist rapid dissolution during soil genesis (Figure 3.2, Markewitz and Richter 2000).

Over time scales of centuries, agricultural land uses can both reduce and renew soil fertility. In the southeastern USA, accelerated erosion degraded many soils, especially from continuous cropping of soils with inadequate vegetative protection and fertility management (Trimble 1974; Daniels 1987). Soil organic matter was lost due to cultivation accelerating decomposition and reducing soil-carbon inputs (Figure 13.1). On the other hand, native soil acidity was greatly ameliorated by liming (Figures 13.5, 13.6). In fact, calcium, nitrogen, phosphorus, and probably magnesium and potassium were enriched by nutrient amendments by long-forgotten farmers (McCracken *et al*. 1989; Richter *et al*. 1994). Such land uses have left a legacy of agriculture in soils of the southeastern USA that will endure for many decades and even centuries.

Over time scales of decades, the Calhoun soil-ecosystem experiment demonstrates that the regrowing pine forest has both depleted and enriched soil fertility, depending on the chemical element. During 40 years of forest development, the upper 0.6 m of mineral soil has substantially acidified (Figure 17.1). Removals of nitrogen, calcium, and magnesium from the mineral soil have outpaced the system's ability to resupply these nutrients in bioavailable form (Richter *et al*. 1994; Markewitz *et al*. 1998). On the other hand, forest regrowth accumulated much organic matter and carbon (Figure 15.1, Table 15.1), not only in aboveground wood and leaves of trees, but also in belowground roots and surface leaf litter. Even still, the rate of reaccumulation of organic matter and carbon in the mineral-soil A horizons (Table 15.1) has been modest (Harrison *et al*. 1995; Richter *et al*. 1999). Mineral-soil supplies of bioavailable potassium and phosphorus are keeping pace with forest demands (Markewitz and Richter 2000), although mineral soil has been greatly depleted in N, and the entire forest ecosystem is slowly reaccumulating this critical nutrient (Table 16.2).

UNDERSTANDING FUTURE SOIL CHANGE: PROFESSOR STRAIN'S BIG QUESTIONS

The analysis of the parcel of land known as the Old Ray Place gives us a small glimpse of the complexity of soil and terrestrial ecosystems that have co-evolved since the Devonian. The analysis also indicates that to improve land management in the foreseeable future, we need a much improved understanding of ecological and historical processes that affect soil change.

The plant ecologist Professor Boyd Strain (e.g., Strain 1985) argued for many years that graduate students research big questions in ecology. One big question is: How can soils and ecosystems best be managed to co-value productivity and environmental quality? This big question can be answered well only if the response is based on technical information provided by soil-ecosystem experiments. Some specific questions raised by our interest in co-valuation of productivity and environmental quality include the following. At what rate and by what mechanisms:

- does the quality of soil organic matter change over time?
- are soils gaining or losing carbon?

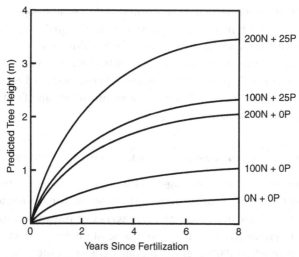

Figure 19.1. Hypothetical dose–response curve for fertilizer response of loblolly pine to nitrogen and phosphorus additions. A strong Nitrogen × Phosphorus interaction is illustrated. A current fertilizer prescription to loblolly pine is 200 kg ha^{-1} of nitrogen and 25 kg ha^{-1} of phosphorus. At least as important as short-term growth response are longer-term effects of inputs on ameliorating low bioavailability of soil nitrogen and phosphorus. The dose–response approach will continue to be the predominant method of management until more detailed understanding is gained about forest-ecosystem nutrition and the fate and effects of fertilizer inputs. Similar conclusions can be stated for all intensive management systems throughout the world.

·

- are soils accumulating additional carbon due to elevated atmospheric CO_2?
- are air pollutants adversely altering biogeochemistry of soils and ecosystems?
- can N_2-fixing symbionts be developed for N-deficient soil?
- can N_2-fixing plants alter the sequestration and quality of soil organic matter?
- does mineral weathering release bioavailable nutrients to plants, microbes, and drainage waters?
- can biodynamic fractions of soil P be managed to promote long-term plant production?
- can we make progress in maximizing nitrogen-use efficiency, increasing nitrogen retention in ecosystems, and minimizing nitrogen loss to drainage water?

A network of soil-ecosystem experiments could be addressing these questions that are fundamentally important to the quality of life on earth in the present and the future. In concluding this book, we propose a network of soil-ecosystem research studies that will focus on soil and ecosystem change. To paraphrase Clements (1916), concerted action is required to organize a network of continuous investigations of soils and the ecosystems of which they are a part.

A PROPOSAL FOR A REGIONAL FOREST-SOILS NETWORK

In the 20th century, forest management greatly intensified on tens of millions of hectares in southeastern North America. The biological potential of these systems is high, and managers and researchers expect net primary productivity of managed southeastern pine forests to increase by as much as 100% over the coming decades. At the turn of the 21st century, a half-million hectares of these forests are being fertilized each year (H.L. Allen, personal communication), mainly with nitrogen and phosphorus, but also with other nutrient elements such as potassium and boron.

These projected increases in net primary productivity in southeastern forests are based in part on dose–response experiments such as those illustrated in Figure 19.1 (p. 211). Only secondarily has research studied the inner workings of nutrients and the detailed ecology of these managed ecosystems (Wells and Jorgensen 1979). For example, the circulation and fate of fertilizer nutrients applied to these forests are not well quantified.

As is true in most new land-management systems, these forestry systems are operating near the margin of our understanding of soil and ecosystem change. Although ecosystem processes that affect soil fertility are understood in southeastern pine forest at least as well as in any other regional forest in the world, significant questions exist about how intensive management will affect soil and ecosystem change over the coming decades. As ecological scientists, we call on stakeholders of the southeastern forest to support an efficient network of soil-ecosystem experiments that will quantify soil and ecosystem change.

A PROPOSAL FOR A GLOBAL SOILS NETWORK

Demands on soil systems continue to rise world-wide (Figures 1.2 to 1.5), yet despite 10 000 years of soil use we have only elementary understanding about the impact of management on soils over time. We call on scientific, financial, conservation, and political institutions to increase substantially support for research that will improve soil management.

We propose the organization of a global network of efficiently run field experiments dedicated to understanding and managing soil and ecosystem change in the foreseeable future. Sites within the network can be selected across a range of soil textures, clay mineralogies, plant species, climates, and management regimes. Siting experiments in locations that encompass a range of ecosystem-controlling variables will greatly benefit both future management and modeling. Since nearly all of the world's on-going soil experiments are currently based in agricultural systems (e.g., Table 4.2), the network of field experiments should explicitly include a wide range of land uses. Since these are studies that accrue value with the passage of time, already-established ecological and natural-resource studies should be reviewed to identify the optimal locations for these soil-ecosystem experiments.

Soil change is not trivial to quantify, and recommendations for how to study soil change can be derived from soil-ecosystem studies such as those at Rothamsted in the UK. Four recommendations that arise from the 40 years of research experience at the Calhoun Forest Experiment are:

1. Soil experiments should be part of larger ecosystem studies. Some of the most significant results from these experiments will likely be ecosystem-level interactions between soil, land use, and climate, and biota (Post *et al.* 1992).
2. Experiments should include manipulative treatments that are relevant to management and to understanding basic system functions. We are currently organizing manipulation treatments at the Calhoun forest that will greatly enrich experimental results in years to come.
3. Permanent plots should be well replicated and sufficiently large in area that they can accommodate repeated within-plot sampling, which must be extensive to ensure that samples are representative.
4. Following each soil collection, samples should be analyzed for ephemeral properties or processes, and then prepared for a sample archive. With the passage of time, the sample archive will become as valuable to the study as the permanent field plots themselves.

INSTITUTIONALIZING A GLOBAL SOILS NETWORK

One approach to instituting a network of long-term soil experiments is to gather collaborative support from national long-term ecological research organizations already established (e.g., in the USA the National Science Foundation's Long-Term Ecological Research program, LTER, and the US Forest Service's Long-Term Site Productivity program, LTSP) as well as from organizations such as the International Long-Term Ecological Research (ILTER) program. Forestry, conservation, environmental, and agricultural organizations of all kinds are potential collaborators and partners in this significant endeavor. Included in this list are the World Bank and the UN's FAO; international agroforestry and forestry research centers such as ICRAF, CIFOR, and CATIE; the International Geographical Union (IGU) and Man and Biosphere (MAB) programs; the International Fertilizer Association; international programs organized to observe global environmental change (e.g., the Inter-American Institute for Global Change Research or IAI); and ecological networks such as those organized by GCTE and the Smithsonian Institution. All these organizations have a stake and an interest in trajectories of soil and ecosystem change.

To quote from a recent ILTER objectives statement:

> Scientific interest in developing long-term ecological research (LTER) programs is expanding very rapidly, reflecting the increased appreciation of their importance in assessing and resolving complex environmental issues. In 1993, the U.S. Long-Term Ecological Research (LTER) Network hosted a meeting on international networking in long-term ecological research. Representatives of scientific programs that focus on ecological research over long temporal and large spatial scales decided there to form the International Long-Term Ecological Research (ILTER) network. They recommended action steps toward developing a worldwide program and the infrastructure necessary to facilitate communication and database management.

The recommended action steps for ILTER all indicate that many natural resource and environmental organizations can support a network of efficiently operated soil and ecosystem experiments. Many of these programs seem well poised to institute a network of soil-ecosystem experiments. For example, the USA's LTER program has recently published a major analytical book for long-term soils studies (Robertson *et al.* 1999).

It can be persuasively argued that many on-going long-term ecological studies cannot fully accomplish their objectives without better quantification of soil change, and a review of these ecological studies indicates important deficiencies in long-term soils research. While much soils research is being conducted, this effort most certainly needs to be greatly expanded and include a more coordinated program of long-term field studies.

The mission for these long-term soil studies can be both understandable to the general public and significant to the scientific community. This mission will be to help ensure and improve management of the world's soils and thereby sustain agricultural, forest, and range systems in the foreseeable future.

The ever-rational King of Brobdingnag would certainly have been one of the strongest supporters of these ambitious proposals to improve our understanding of soil change.

Epilogue

God will not ask thee thy race,
Nor thy birth.
Alone he will ask of thee,
What hast thou done with the land I gave thee?

Persian proverb

Recommended readings

This book is read chapter by chapter in a graduate class at Duke University in conjunction with at least some of the following readings.

Chapter

1   Perlin, J. 1989. *A Forest Journey*. Harvard University Press, Cambridge, MA. (Chapters 1 and 2)

2   Sánchez, P.A. 1994. Tropical soil fertility research: Towards a second paradigm. Pp. 65–88 in *Transactions of 15th World Congress of Soil Science*, Vol. 1. Acapulco, Mexico.

3   Burke, I., W.K. Lauenroth, and C.A. Wessman. 1998. Progress in understanding biogeochemical cycles at regional to global scales. Pp. 165–194 in M.L. Mace and P.M. Groffman (eds.) *Successes, Limitations, and Frontiers in Ecosystem Science*. Springer-Verlag, New York.

4   Pickett, S.T.A. 1989. Space-for-time substitution as an alternative to long-term studies. Pp. 110–135 in G.E. Likens (ed.) *Long-term Studies in Ecology: Approaches and Alternatives*. Springer-Verlag, New York.

   Jenkinson, D.S. 1991. The Rothamsted classical experiments: Are they still of use? *Agronomy Journal* 83: 2–10.

5   Barnes, B.V., D.R. Zak, S.R. Denton, and S.H. Spurr. 1998. *Forest Ecology*. 4th edn. Wiley, New York. (Chapter 3)

6   Christensen, N.L. 1989. Landscape history and ecological change. *Journal of Forest History Society* 33: 116–124.

   Likens, G.E., C.T. Driscoll, and D.C. Buso. 1992. Long-term effects of acid rain: Response and recovery of a forest ecosystem. *Science* 272: 244–246.

7   Retallack, G.J. 1990. *Soils of the Past*. Unwin Hyman, London. (Chapters 1 and 2)

   Jackson, W. 1992. Living soil: *Orion* editorial. *Orion* 11(2): 3.

   Chadwick, O.A., L.A. Derry, P.M. Vitousek, B.J. Huebert, and L.O.

Hedin. 1999. Changing sources of nutrients during four million years of ecosystem development. *Nature* 397: 491–497.

8 Buol, S.A., F.D. Hole, and R.J. McCracken. 1989. *Soil Genesis and Classification.* 3rd edn. University of Iowa Press, Ames. (Chapter 17, Ultisols)

9 Eswaran, H. and W.C. Bin. 1978. A study of deep weathering profile on granite in peninsular Malaysia: I. Physiochemical and micromorphological properties. *Soil Science Society of America Journal* 42: 144–149.

10 Brimhall, G.H., O.A. Chadwick, C.J. Lewis *et al.* 1991. Deformational mass transport and invasive processes in soil evolution. *Science* 255: 695–702.

11 Smith, B.D. 1989. Origins of agriculture in eastern North America. *Science* 246: 1566–1571.

12 Trimble, S.W. 1974. *Man-induced Soil Erosion on the Southern Piedmont, 1700–1970.* Soil Conservation Society, Ankeny, IA.

13 McCracken, R.J., R.B. Daniels, and W.E. Fulcher. 1989. Undisturbed soils, landscapes, and vegetation in a North Carolina Piedmont virgin forest. *Soil Science Society of America Journal* 53: 1146–1152.

14 Cowdrey, A.E. 1996. *This Land, This South.* University Press of Kentucky, Lexington. (Chapter 9)
Page, G. 1968. Some effects of coniferous crops on soil properties. *Commonwealth Forestry Review* 47: 52–62.

15 Torn, M.S., S.E. Trumbore, O.A. Chadwick, P.M. Vitousek, and D.M. Hendricks. 1997. Mineral control of soil carbon storage and turnover. *Nature* 389: 170–173.
Jorgensen, J., C.G. Wells and L.J. Metz. 1980. Nutrient changes in decomposing loblolly pine forest floor. *Soil Science Society of America Journal* 44: 1307–1314.

16 Jenkinson, D.S. 1971. The accumulation of nitrogen in soil left uncultivated. *Report of the Rothamsted Experiment Station 1970*, 2: 113–137.

17 Johnston, A.E., K.W.T. Goulding, and P.R. Poulton. 1986. Soil acidification during more than 100 years under permanent grassland and woodland at Rothamsted. *Soil Use and Management* 2: 3–10.

18 Schmidt, J.P., S.W. Buol, and E.J. Kamprath. 1997. Soil phosphorus dynamics during 17 years of continuous cultivation: A method to estimate long-term P availability. *Geoderma* 78: 59–70.
Newman, E.I. 1997. Phosphorus balance of contrasting farming systems: Past and present. Can food production be sustainable? *Journal of Applied Ecology* 34: 1334–1347.

Carbonic acid weathering reactions

220   Fundamental chemical reactions of the carbonic acid weathering system
include the following (Stumm and Morgan 1981; Richter and Markewitz
1995b). In the soil, where soil respiration increases partial pressures of
carbon dioxide:

$$CO_{2(g)} + H_2O \rightarrow H_2CO_3{}^*$$

that is mathematically stated:

$$(pCO_2)\,(K_H) = H_2CO_3{}^*$$

where pCO_2 is the partial pressure of CO_2, K_H is Henry's constant ($10^{-1.32}$,
at 15 °C and low ionic strength), and $H_2CO_3{}^*$ is the sum of dissolved and
hydrated CO_2. In turn, $H_2CO_3{}^*$ dissociates to H^+, $HCO_3{}^-$, and potentially
$CO_3{}^{2-}$:

$$H_2CO_3{}^* \rightarrow H^+ + HCO_3{}^-$$
$$HCO_3{}^- \rightarrow H^+ + CO_3{}^{2-}$$

that are mathematically stated:

$$[H^+]\,[HCO_3{}^-] / [H_2CO_3{}^*] = K_1$$
$$[H^+]\,[CO_3{}^{2-}] / [HCO_3{}^-] = K_2$$

where K_1 is the first dissociation constant of carbonic acid ($10^{-6.419}$ at
15 °C), and K_2 is the second ($10^{-10.33}$).

APPENDIX II

Simulation of bomb-produced ^{14}C in the forest floor at the Calhoun Experimental Forest, SC

The following model and parameters simulate incorporation and decomposition of ^{14}C in an aggrading forest floor at the Calhoun Experimental Forest, SC. Litterfall estimates are modified from Urrego's (1993)
measurements at the Calhoun forest, combined with litterfall data from
a series of loblolly pine stands of different ages, summarized in Switzer
and Nelson (1972). Estimates for atmospheric Δ^{14}C-CO$_2$ are taken from
Burchuladze et al. (1989). Decomposition coefficients for the simulation
model come from the litter-sandwich experiment of Jorgensen et al.
(1980) performed in a loblolly pine stand growing on an Appling soil in
Duke University Forest, NC:

$$C_{Ohor} = C_i \left[\exp \left(4.67 - 0.561 \left(\ln \text{year } i + 1 \right) \right) \right]$$

where C_{Ohor} is carbon of a litterfall cohort remaining in the O horizon in
kg ha^{-1}, C_i is carbon in litterfall in year i in kg ha^{-1}, and year i is the age in
years of the litterfall.

Year	Simulated litterfall C (kg ha^{-1} year^{-1})	Atmosphere Δ^{14}C-CO$_2$ (per mil)	Simulated Δ^{14}C in O horizon
1950	—	− 14.2	—
1951	—	− 43.7	—
1952	—	− 13.7	—
1953	—	− 25.3	—
1954	—	− 6.7	—
1955	—	16.6	—
1956	—	69.5	—
1957	—	88.7	—
1958	250	255.0	255.0
1959	350	290.0	278.1
1960	450	255.0	265.5

(cont.)

Year	Simulated litterfall C (kg ha^{-1} year^{-1})	Atmosphere Δ^{14}C-CO_2 (per mil)	Simulated Δ^{14}C in O horizon
1961	650	267.0	266.4
1962	900	365.2	313.1
1963	1200	823.1	538.2
1964	1600	891.6	678.4
1965	2000	764.8	700.1
1966	2400	626.8	664.3
1967	2800	597.1	638.2
1968	3200	551.5	609.6
1969	3500	510.3	581.0
1970	3500	577.7	583.7
1971	3500	534.0	573.2
1972	3500	532.8	566.2
1973	3500	426.3	538.6
1974	3000	397.7	518.6
1975	3000	368.0	498.9
1976	3000	332.0	478.3
1977	3000	297.7	458.8
1978	3000	344.9	449.9
1979	3000	294.5	434.2
1980	3000	279.3	419.8
1981	3000	271.5	407.2
1982	3000	265.7	396.2
1983	3000	229.8	382.2
1984	3000	215.9	369.2
1985	3000	208.7	357.7
1986	3000	189.8	345.8
1987	3000	167.9	333.2
1988	3000	170.9	323.3
1989	3000	162.9	313.8
1990	3000	154.9	304.8
1991	3000	146.9	296.0
1992	3000	138.9	287.6
1993	3000	130.9	279.5
1994	3000	122.9	271.6
1995	3000	114.9	263.9
1996	3000	106.9	256.3

Appendix III

Sources of variation in the Calhoun Experimental Forest's main analysis of variance (ANOVA)

To test soil change in soil phosphorus and other nutrients, the following generalized ANOVA was used for the field experiment at the Calhoun Experimental Forest, SC. The mean square error of Block × Year effects was used to test significance of the main Year effects. The ANOVA that tested change in other soil nutrients was similar, except that up to 6 degrees of freedom were used to test main effects of time due to the seven repeated samplings between 1962 and 1997.

Source of variation	Degrees of freedom
Block	3
Spacing	1
Block × Spacing	3
Year (1962 vs. 1990)	1
Block × Year	3
Spacing × Year	1
Error	3
Total	15

Total elemental concentrations for soils from the Calhoun Experimental Forest, SC

224   Soil profiles were collected from the four experimental blocks of the long-term field study (site P-1) between 1990 and 1994. Rock samples were collected from the bottom of a gully system along nearby Sparks Creek. Table 6.1 shows data for chemical elements that are found in greater concentration.

Depth (m)	Horizon	Mn	P	Ba	Zr	Sr	V	Cr	Cu	Ni	Sc
						(μg g^{-1})					
0–0.15	A + E	217	218	141	392	31	11	17	4	1	1
0.15–0.35	E	216	228	157	417	46	25	16	6	8	2
0.35–0.6	EB	172	327	177	407	48	73	39	15	21	5
0.6–1.0	B	152	419	141	324	53	107	50	22	32	8
1.0–1.5	B	164	387	153	306	57	100	43	22	28	9
2.0–2.5	B	210	343	303	272	76	70	24	23	20	7
3.0–3.5	BC	287	498	423	238	130	65	21	77	11	9
4.0–4.5	CB	355	417	705	264	201	47	11	43	8	5
5.0–5.5	C	367	327	675	244	193	43	10	82	11	5
6.0–7.0	C	344	233	595	181	247	43	17	43	13	4
7.0–8.0	C	352	263	627	246	295	40	15	26	10	3
Rock	R	556	496	157	116	499	44	19	3	8	4
					Coefficients of variation (%)						
0–0.15		30	4	8	15	10	8	0	18	53	40
0.15–0.35		30	10	8	11	5	34	23	19	33	28
0.35–0.6		3	12	24	16	7	16	12	3	33	23
0.6–1.0		12	8	26	46	7	25	1	23	37	28
1.0–1.5		32	5	9	18	1	13	7	15	67	28
2.0–2.5		21	13	35	11	24	7	28	39	1	11
3.0–3.5		13	2	42	42	16	13	33	57	13	5
4.0–4.5		23	24	22	11	33	4	20	22	8	2
5.0–5.5		28	10	1	42	32	28	22	91	54	34
6.0–7.0		57	1	47	8	31	36	4	74	13	52
7.0–8.0		39	25	48	17	36	10	25	41	23	85
Rock		9	2	13	14	1	1	3	9	12	8

References

226

Adams, F. (ed.) 1984. *Soil Acidity and Liming*. 2nd edn. American Society of Agronomy, Madison, WI.

Ågren, G.I. and E. Bosatta. 1996. *Theoretical Ecosystem Ecology: Understanding Element Cycles*. Cambridge University Press, New York.

Alban, D.H. 1982. Effects of nutrient accumulation by aspen, spruce, and pine on soil properties. *Soil Science Society of America Journal* 46: 853–861.

Alexiades, C.A., N.A. Polyzopoulos, N.S. Koroxenides, and G.S. Axaris. 1973. High trioctahedral vermiculite content in the sand, silt, and clay fractions of a gray brown podzolic soil in Greece. *Soil Science* 116: 363–375.

Algeo, T.J., R.A. Berner, J.B. Maynard, and S.E. Scheckler. 1995. Late Devonian oceanic anoxic events and biotic crises: "Rooted" in the evolution of vascular plants? *GSA Today* 5: 44–66.

Allison, F.E. 1955. The enigma of soil nitrogen balance sheets. *Advances in Agronomy* 7: 213–250.

Alriksson, A. 1998. *Afforestation of Farmland*. Acta Universitatis Agriculturae Sueciae, Silvestria 57. Uppsala, Sweden.

Ambrose, S. 1987. Chemical and isotopic techniques of diet reconstruction in eastern North America. Pp. 87–107 in W. Keegan (ed.) *Emergent Horticultural Economics of the Eastern Woodlands*. Southern Illinois University Press, Carbondale.

Amundson, R.G. and E.A. Davidson. 1990. Carbon dioxide and nitrogenous gases in the soil atmosphere. *Journal of Geophysical Exploration* 38: 13–41.

Anderson, D.W. 1991. Long-term ecological research: A pedological perspective. Pp. 115–134 in P.G. Risser (ed.) *Long-term Ecological Research*. Scientific Committee on Problems of the Environment (SCOPE) of the International Council of Scientific Unions (IUCN). Wiley, New York.

Antonovics, J., A.D. Bradshaw, and R.G. Turner. 1971. Heavy metal tolerance in plants. *Advances in Ecological Research* 7: 1–85.

April, R. and D. Keller. 1990. Mineralogy of the rhizosphere in forest soils of the eastern United States. *Biogeochemistry* 9: 1–18.

April, R. and R. Newton. 1992. Mineralogy and mineral weathering. Pp. 378–425 in D.W. Johnson and S.E. Lindberg (eds.) *Atmospheric Deposition and Nutrient Cycling in Forest Ecosystems*. Springer-Verlag, New York.

Asch, D. and N. Asch. 1985a. Prehistoric plant cultivation in west-central Illinois. Pp. 149–203 in R.I. Ford (ed.) *Prehistoric Food Production in North America*. Anthropological Papers 75. Museum of Anthropology, University of Michigan, Ann Arbor.

Asch, D. and N. Asch. 1985b. Archeobotany. Pp. 322–401 in B. Stafford and M. Sant (eds.) *Smiling Dan: Structure and Function at a Middle Woodland Settlement in the Illinois Valley*. Kampsville Archeological Center Research Series 2. Kampsville Archeological Center, IL.

Barber, S.A. 1979. Soil phosphorus after 25 years of cropping with five rates of phosphorus application. *Communications in Soil Science and Plant Analysis* 10: 1459–1468.

Barber, B.L. and D.H. Van Lear. 1984. Weight loss and nutrient dynamics in decomposing woody loblolly pine slash. *Soil Science Society of America Journal* 48: 906–910.

Barnes, B.V., D.R. Zak, S.R. Denton, and S.H. Spurr. 1998. *Forest Ecology*. 4th edn. Wiley, New York.

Barnhisel, R.I. and P.M. Bertsch. 1989. Chlorites and hydroxy-interlayered vermiculite and smectite. Pp. 729–779 in J.B. Dixon and S.B. Weed (eds.) *Minerals in Soil Environments*. 2nd edn. Soil Science Society of America, Madison, WI.

Beck, M.A. and P.A. Sánchez. 1994. Soil phosphorus fraction dynamics during 18 years of cultivation on a Typic Paleudult. *Soil Science Society of America Journal* 58: 1424–1431.

Beerbower, R. 1985. Early development of continental ecosystems. Pp. 47–91 in B.H. Tiffney (ed.) *Geological Factors and the Evolution of Plants*. Yale University Press, New Haven, CT.

Behrensmeyer, A.K., J.D. Damuth, W.A. DiMichele, R. Potts, H.-D. Sues, and S.L. Wing (eds.) 1992. *Terrestrial Ecosystems through Time*. University of Chicago Press, IL.

Berner, R.A. 1995. Chemical weathering and its effect on atmospheric CO_2 and climate. *Reviews in Mineralogy* 31: 565–583.

Berner, R.A. and K.A. Maasch. 1996. Chemical weathering and controls on O_2 and CO_2: Fundamental principles were enunciated by J.J. Ebelmen in 1845. *Geochimica et Cosmochimica Acta* 60: 1633–1637.

Berry, W. 1995. *Another Turn of the Screw*. Counterpoint, Washington, DC.

Berthelin, J. and C. Leyval. 1982. Ability of symbiotic and non-symbiotic rhizospheric microflora of maize (*Zea mays*) to weather micas and to promote plant growth and plant nutrition. *Plant and Soil* 68: 369–377.

Berthelin, J., M. Bonne, G. Belgy, and F.X. Wedraogo. 1985. A major role for nitrification in the weathering of minerals in brown acid forest soils. *Geomicrobiology Journal* 4: 175–190.

Billings, W.D. 1938. The structure and development of old-field short-leaf pine stands and certain associated physical properties of the soil. *Ecological Monographs* 8: 437–499.

Binkley, D. 1986. *Forest Nutrition Management*. Wiley, New York.

Binkley, D. and C. Giardina. 1997. Why do trees affect soils? The warp and woof of tree–soil interactions. *Biogeochemistry* 16: 1–18.

Binkley, D. and D.D. Richter. 1987. Nutrient cycles and H^+ budgets of forest ecosystems. *Advances in Ecological Research* 16: 1–51.

Binkley, D., D. Valentine, C. Wells, and U. Valentine. 1989. An empirical analysis of the factors contributing to 20-yr decrease in soil pH in an old-field plantation of loblolly pine. *Biogeochemistry* 8: 39–54.

Binns, J.A. 1803. *Treatise on Practical Farming*. Frederick-town, MD.

Bischof, K.G. 1847. *Lehrbuch der Chemischen und Physikalischen Geologie*, Vol. 1. Marcus. (Publ. transl. 1854 by Cavendish Society, London.)

Black, C.A. 1968. *Soil–Plant Relationships*. 2nd edn. Wiley, New York.

Blackmer, A.M., T.F. Morris, D.R. Keeney, R.D. Voss, and R. Killorn. 1991.

Estimating Nitrogen Needs for Corn by Soil Testing: Iowa 1991. Iowa State University Extension Pamphlet 1381. Cooperative Extension Service, Ames, IA.

Bormann, B.T., F.H. Bormann, W.B. Bowden *et al.* 1993. Rapid N_2 fixation in pines, alder, and locust: Evidence from the sandbox ecosystem study. *Ecology* 74: 583–598.

Bornemisza, E. 1982. Nitrogen cycling in coffee plantations. *Plant and Soil* 67: 241–246.

Bourne, E.G. 1904. *Narratives of the Career of Hernando de Soto.* A.S. Barnes and Co., New York.

Boyle, J.R. and G.K. Voigt. 1973. Biological weathering of silicate minerals. Implications for tree nutrition and soil genesis. *Plant and Soil* 38: 191–201.

Bradshaw, A.D. and M.J. Chadwick. 1980. *The Restoration of Land.* Blackwell, Oxford.

Brady, N.C. and R. Weil. 1996. *The Nature and Properties of Soil.* 11th edn. Prentice-Hall, Upper Saddle River, NJ.

Bray, R.H. and L.T. Kurtz. 1945. Determination of total, organic, and available forms of phosphorus in soils. *Soil Science* 59: 39–45.

Brimhall, G.H., O.A. Chadwick, C.J. Lewis *et al.* 1991. Deformational mass transport and invasive processes in soil evolution. *Science* 255: 695–702.

Brown, J.R. 1994. The Sanborn field experiment. Pp. 39–52 in R.A. Leigh and A.E. Johnston (eds.) *Long-term Experiments in Agricultural and Ecological Sciences.* CAB International, Wallingford, UK.

Buford, M.A. 1991. Performance of four yield models for predicting stand dynamics of a 30-year-old loblolly pine (*Pinus taeda* L.) spacing study. *Forest Ecology and Management* 46: 23–38.

Buol, S.A. 1994. Saprolite-regolith taxonomy. Pp. 119–132 in D.L. Cremeens, R.B. Brown, and J.H. Huddleston (eds.) *Whole Regolith Pedology.* SSSA Special Publication 34. Soil Science Society of America, Madison, WI.

Buol, S.A. 1995. Sustainability of soil use. *Annual Reviews of Ecology and Systematics* 26: 25–44.

Buol, S.A. and S.B. Weed. 1991. Saprolite–soil transformations in the Piedmont and Mountains of North Carolina. *Geoderma* 51: 15–28.

Buol, S.A., F.D. Hole, and R.J. McCracken. 1989. *Soil Genesis and Classification.* 3rd edn. University of Iowa Press, Ames.

Burchuladze, A.A., M. Chudy, I.V. Eristavi *et al.* 1989. Anthropogenic ^{14}C variations in atmospheric CO_2 and wines. *Radiocarbon* 31: 771–776.

Burke, I., W.K. Lauenroth, and C.A. Wessman. 1998. Progress in understanding biogeochemical cycles at regional to global scales. Pp. 165–194 in M.L. Mace and P.M. Groffman (eds.) *Successes, Limitations, and Frontiers in Ecosystem Science.* Springer-Verlag, New York.

Burke, I.C., W.K. Lauenroth, and D.P. Coffin. 1995. Soil organic matter recovery in semiarid grasslands: Implications for the conservation reserve program. *Ecological Applications* 5: 793–801.

Calvert, C.S., S.W. Buol, and S.B. Weed. 1980. Mineralogical characteristics and transformations of a vertical rock–saprolite–soil sequence in the North Carolina Piedmont: I. Profile morphology, chemical composition, and mineralogy. *Soil Science Society of America Journal* 44: 1096–1103.

Carroll, D. 1970. *Rock Weathering.* Plenum Press, New York.

Carter, M.R. (ed.) 1993. *Soil Sampling and Methods of Analysis.* Canadian Society of Soil Science/Lewis Publishers, Boca Raton, FL.

Cassman, K.G. and P.L. Pingali. 1994. Extrapolating trends from long-term experiments to farmer's fields: The case for irrigated rice systems in Asia.

Pp. 63–84 in V. Barnet, R. Payne, and R. Steiner (eds.) *Agricultural Sustainability: Economic, Environmental and Statistical Considerations.* Wiley, Chichester, UK.

Cassman, K.G., S.K. DeDatta, D.C. Olk *et al.* 1995. Yield decline and the nitrogen economy of long-term experiments on continuous, irrigated rice systems in the tropics. Pp. 181–222 in R. Lal and B.A. Stewart (eds.) *Soil Management: Experimental Basis for Sustainability and Environmental Quality.* CRC Lewis Publishers, Boca Raton, FL.

Chadwick, O.A., G.H. Brimhall, and D.M. Hendricks. 1990. From a black to a gray box – A mass balance interpretation of pedogenesis. *Geomorphology* 3: 369–390.

Chadwick, O.A., L.A. Derry, P.M. Vitousek, B.J. Huebert, and L.O. Hedin. 1999. Changing sources of nutrients during four million years of ecosystem development. *Nature* 397: 491–497.

Chang, S.C. and M.L. Jackson. 1957. Fractionation of soil phosphorus. *Soil Science* 84: 133–144.

Chapin, F.S., L.R. Walker, C.L. Fastie, and L.C. Sarman. 1994. Mechanisms of primary succession following deglaciation at Glacier Bay, Alaska. *Ecological Monographs* 64: 149–175.

Charles, A.D. 1987. *The Narrative History of Union County South Carolina.* Reprint Company Pub., Spartanburg, SC.

Chizhikov, P.N. 1968. The lower boundary of soil. *Soviet Soil Science* 11: 1489–1493.

Christensen, B.T., J. Petersen, V. Kjellerup, and U. Trentemøller. 1994. The Askov long-term experiments on animal manure and mineral fertilizers: 1894–1994. SP Report 43, Danish Institute of Plant and Soil Science. Skovbrynet, Lyngby.

Christensen, N.L. 1989. Landscape history and ecological change. *Journal of Forest History Society* 33: 116–124.

Clark, T. 1968. *The Emerging South.* Oxford University Press, New York.

Clements, F.E. 1916. *Plant Succession.* Publication 242. Carnegie Institution of Washington, Washington, DC.

Cochrane, W.W. 1979. *The Development of American Agriculture.* University of Minnesota Press, Minneapolis.

Coile, T.S. 1940. *Soil Changes Associated with Loblolly Pine Succession on Abandoned Agricultural Land on the Piedmont Plateau.* Bulletin 5. School of Forestry, Duke University, Durham, NC.

Cole, D.W. and M. Rapp. 1982. Elemental cycling in forest ecosystems. Pp. 341–409 in D.E. Reichle (ed.) *Dynamics of Forest Ecosystems.* Cambridge University Press, Malta.

Comerford, N.B., W.G. Harris, and D. Lucas. 1990. Release of non-exchangeable potassium from a highly weathered forested Quartzipsamment. *Soil Science Society of America Journal* 54: 1421–1426.

Cope, J.T. 1981. Effects of 50 years of fertilization with phosphorus and potassium on soil test levels and yields at six locations. *Soil Science Society of America Journal* 45: 342–347.

Cosby, B.J., G.M. Hornberger, J.N. Galloway, and R.F. Wright. 1985. Modeling the effects of acid deposition: Assessment of a lumped parameter model of soil water and stream water chemistry. *Water Resources Research* 21: 51–63.

Cowdrey, A.E. 1996. *This Land, This South.* University Press of Kentucky, Lexington.

Cox, F.R., E.J. Kamprath, and R.E. McCollum. 1981. A descriptive model of soil test nutrient levels following fertilization. *Soil Science Society of America Journal* 45: 529–532.

Creemans, D.L., R.B. Brown, and J.H. Huddleston (eds.) 1994. *Whole Regolith Pedology.* SSSA Special Publication 34. Soil Science Society of America, Madison, WI.

Crews, T.E., K. Kitayama, J.H. Fownes *et al.* 1995. Changes in soil phosphorus fractions and ecosystem dynamics across a long chronosequence in Hawaii. *Ecology* 76: 1407–1424.

Cronan, C.S., W.A. Reiners, R.C. Reynolds, and G.E. Lang. 1978. Forest floor leaching: Contributions from mineral, organic, and carbonic acids in New Hampshire subalpine forests. *Science* 200: 309–311.

Cronon, W. 1983. *Changes in the Land.* Hill and Wang, New York.

Cross, A.F. and W.H. Schlesinger. 1994. A literature review and evaluation of the Hedley fractionation: Applications to the biogeochemical cycle of soil phosphorus in natural ecosystems. *Geoderma* 64: 197–214.

Dalal, R.C. 1977. Soil organic phosphorus. *Advances in Agronomy* 29: 83–113.

Daniels, R.B. 1987. Soil erosion and degradation in the southern Piedmont of the USA. Pp. 407–428 in M.G. Wolman and F.G.A. Fournier (eds.) *Land Transformation in Agriculture.* Wiley, London.

Daniels, R.B. and L.A. Nelson. 1987. Soil variability and productivity: Future developments. Pp. 279–291 in *Future Developments in Soil Science Research.* Soil Science Society of America, Madison, WI.

Darwin, C. 1897. *The Formation of Vegetable Mold, Through the Action of Worms.* D. Appleton, New York.

DeBell, D.S., W.R. Harms, and C.G. Whitesell. 1989. Stockability: A major factor in productivity differences between *Pinus taeda* plantations in Hawaii and the southeastern United States. *Forest Science* 35: 708–719.

Delcourt, H.R. and W.F. Harris. 1980. Carbon budget of the southern U.S. biota: Analysis of historical change in trend from source to sink. *Science* 210: 321–323.

Delcourt, P.A., H.R. Delcourt, D.F. Morse, and P.A. Morse. 1993. History, evolution, and organization of vegetation and human culture. Pp. 47–79 in W.H. Martin, S.G. Boyce, and A.C. Esternacht (eds.) *Biodiversity of the Southeastern United States.* Wiley, New York.

Dickson, B.A. and R.L. Crocker. 1953. A chronosequence of soils and vegetation near Mt. Shasta, California. *Journal of Soil Science* 4: 142–154.

Dobyns, H.F. 1983. *Their Number Become Thinned.* University of Tennessee Press, Knoxville.

Douglas, L.A. 1989. Vermiculites. Pp. 635–674 in J.B. Dixon and S.B. Weed (eds.) *Minerals in Soil Environments.* 2nd edn. Soil Science Society of America, Madison, WI.

Driscoll, C.T. and G.E. Likens. 1982. Hydrogen ion budget of an aggrading forest ecosystem. *Tellus* 34: 283–292.

Driscoll, C.T., G.E. Likens, L.O. Hedin, J.S. Eaton, and F.H. Bormann. 1989. Changes in the chemistry of surface waters. *Environmental Science and Technology* 23: 137–143.

Duchaufour, P. 1982. *Pedology.* Allen and Unwin, London.

Dunscomb, J. 1992. *Effects of Land Use on Organic Matter in some Piedmont Soils.* Masters Project, School of the Environment, Duke University, Durham, NC.

Dyke, G.V. 1991. *John Bennett Lawes: The Record of his Genius.* Wiley, Chichester, UK.

Earle, C. 1992. *Geographical Inquiry and American Historical Problems.* Stanford University Press, Stanford, CA.

Ebelmen, J.J. 1845. Sur les produits de la décomposition des espèces minérales de la famille des silicates. *Annuales des Mines* 7: 3–66.

Ebermayer, E.W.H. 1876. *Die Gesammte Lehre der Waldstreu, mit Rücksicht auf die Chemische Statik des Waldbaues.* J. Springer, Berlin.

Environmental Protection Agency. 1971. *Methods of Chemical Analysis for Water and Wastes.* US Environmental Protection Agency, Cincinnati, OH.

Eswaran, H. and W.C. Bin. 1978. A study of deep weathering profile on granite in peninsular Malaysia: I. Physiochemical and micromorphological properties. *Soil Science Society of America Journal* 42: 144–149.

Evans, F. 1956. Ecosystem as the basic unit in ecology. *Science* 123: 1127–1128.

Evans, J. 1992. *Plantation Forestry in the Tropics.* 2nd edn. Clarendon Press, Oxford.

Evans, J. 1994. Long-term experimentation in forestry and site change. Pp. 83–94 in R.A. Leigh and A.E. Johnston (eds.) *Long-term Experiments in Agricultural and Ecological Sciences.* CAB International, Wallingford, UK.

Fanning, D.S. and M.C.B. Fanning. 1989. *Soil Morphology, Genesis, and Classification.* Wiley, New York.

Fanning, D.S., V.Z. Keramidas, and M.A. El-Desoky. 1989. Micas. Pp. 195–258 in J.B. Dixon and S.B. Weed (eds.) *Minerals in Soil Environments.* 2nd edn. Soil Science Society of America, Madison, WI.

FAO–UNESCO. 1971. *Soil Map of the World, 1: 5,000,000. Vol. IV: South America.* United Nations Educational, Scientific and Cultural Organization, Paris.

FAO–UNESCO. 1974. *Soil Map of the World, 1:5,000,000. Vol. II: Legend.* United Nations Educational, Scientific and Cultural Organization, Paris.

FAO–UNESCO. 1975. *Soil Map of the World, 1:5,000,000. Vol. III: Mexico and Central America.* United Nations Educational, Scientific and Cultural Organization, Paris.

FAO–UNESCO. 1977a. *Soil Map of the World, 1:5,000,000. Vol. VI: Africa.* United Nations Educational, Scientific and Cultural Organization, Paris.

FAO–UNESCO. 1977b. *Soil Map of the World, 1:5,000,000. Vol. VII: South Asia.* United Nations Educational, Scientific and Cultural Organization, Paris.

FAO–UNESCO. 1979. *Soil Map of the World, 1:5,000,000. Vol. IX: Southeast Asia.* United Nations Educational, Scientific and Cultural Organization, Paris.

FAO–UNESCO. 1988. *Soil Map of the World, Revised Legend.* United Nations Educational, Scientific and Cultural Organization, Paris.

Federer, C.A., J.W. Hornbeck, L.M. Tritton, C.W. Martin, R.S. Pierce, and C.T. Smith. 1989. Long term depletion of calcium and other nutrients in eastern U.S. forests. *Environmental Management* 13: 593–601.

Fisher, R.A. 1951. *The Design of Experiments.* 6th edn. Hafner, New York.

Fisher, R.F. 1990. Amelioration of soils by trees. Pp. 290–300 in S.P. Gessel (ed.) *Sustained Productivity of Forest Soils.* Proceedings of the Seventh North American Forest Soils Conference, Vancouver, British Columbia, Canada.

Foster, D.R. 1999. *Thoreau's Country.* Harvard University Press, Cambridge, MA.

Fox, R.L. and E.J. Kamprath. 1970. Phosphorus sorption isotherms for evaluating the phosphate requirements of soils. *Soil Science Society of America Proceedings* 34: 902–907.

Fox, T.R. and N.B. Comerford. 1990. Low-molecular-weight organic acids in selected forest soils of the southeastern USA. *Soil Science Society of America Proceedings* 54: 1139–1144.

Foy, C.D. 1984. Physiological effects of hydrogen, aluminum, and manganese toxicities in acid soils. Pp. 57–69 in F. Adams (ed.) *Soil Acidity and Liming*. 2nd edn. American Society of Agronomy, Madison, WI.

Franzmeier, D.P. and E.P. Whiteside. 1963. *A Chronosequence of Podzols in Northern Michigan*. Quarterly Bulletin 46. Michigan Agricultural Experiment Station, East Lansing.

Fulmer, J.L. 1950. *Agricultural Progress in the Cotton Belt since 1920*. University of North Carolina Press, Chapel Hill.

Fundaburk, E.L. 1958. *Southeastern Indians: Life Portraits*. E.L. Fundaburk Publ., Luverne, AL.

Galloway, J.N., H. Levy, and P.S. Kasibhatla. 1994. Year 2020: Consequences of population growth and development on deposition of oxidised nitrogen. *Ambio* 23: 120–123.

Gandhi, M.K. 1940. *An Autobiography, or, the Story of My Experiments with Truth*. Navajivan Publishing House, Ahmedabad.

Gardner, R.A. 1967. *Sequence of Podsolic Soils along the Coast of Northern California*. PhD Thesis, University of California, Berkeley.

Gensel, P.G. and H.N. Andrews. 1984. *Plant Life in the Devonian*. Praeger, New York.

Gnau, C.B. 1992. *Modeling the Hydrologic Cycle during 25 Years of Forest Development*. Masters Project, School of the Environment, Duke University, Durham, NC.

Gobran, G.R. and S. Clegg. 1996. A conceptual model for nutrient availability in the mineral soil-root system. *Canadian Journal of Soil Science* 76: 125–131.

Goh, K.M. and L.M. Condron. 1989. Plant availability of phosphorus accumulated from long-term applications of superphosphate and effluent to irrigated pastures. *New Zealand Journal of Agricultural Research* 32: 45–51.

Gorham, E. 1991. Biogeochemistry: Its origins and development. *Biogeochemistry* 13: 199–239.

Grace, P. and J.M. Oades. 1994. Long-term field trials in Australia. Pp. 53–81 in R.A. Leigh and A.E. Johnston (eds.) *Long-term Experiments in Agricultural and Ecological Sciences*. CAB International, Wallingford, UK.

Graham, R.C., K.R. Tice, and W.R. Guertal. 1994. The pedogenic nature of weathered rock. Pp. 21–40 in D.L. Creemans, R.B. Brown, and J.H. Huddleston (eds.) *Whole Regolith Pedology*. SSSA Special Publication 34. Soil Science Society of America, Madison, WI.

Gray, D.H. and A.T. Leiser. 1989. *Biotechnical Slope Protection and Erosion Control*. Krieger Publishing Co., Malabar, FL.

Gray, L.C. 1933. *History of Agriculture in the Southern United States to 1860*. Carnegie Institution of Washington, Washington, DC.

Greenland, D.J. 1994. Long-term cropping experiments in developing countries: the need, the history, and the future. Pp. 187–209 in R.A. Leigh and A.E. Johnston (eds.) *Long-term Experiments in Agricultural and Ecological Sciences*. CAB International, Wallingford, UK.

Greenland, D.J. and I. Szabolcs (eds.) 1994. *Soil Resilience and Sustainable Land Use*. CAB International, Wallingford, UK.

Gupta, A.P., R.P. Narwal, R.S. Antil, and S. Dev. 1992. Sustaining soil fertility with organic-C, N, P, and K by using farmyard manure and fertility-N in a semiarid zone: A long-term study. *Arid Soil Research and Rehabilitation* 6: 243–251.

Haagsma, T. and M.H. Miller. 1963. The release of non-exchangeable soil potassium to cation-exchange resins as influenced by temperature, moisture and exchanging ion. *Soil Science Society of America Proceedings* 27: 153–156.

Hack, J.T. 1960. Interpretation of erosional topography in humid climates of temperate regions. *American Journal of Science* 258A: 80–97.

Hall, A.R. 1948. *Soil Erosion and Agriculture in the Southern Piedmont: A History.* PhD Dissertation, Duke University, Durham, NC.

Hall, B. 1829. *Forty Etchings from Sketches made with the Camera Lucida in North America in 1827 and 1828.* Cadell, Edinburgh.

Harden, J.W. 1988. Genetic interpretations of elemental and chemical differences in a soil chronosequence, California. *Geoderma* 43: 179–193.

Harden, J.W., E.T. Sundquist, R.F. Stallard, and R.K. Mark. 1992. Dynamics of soil carbon during deglaciation of the Laurentide Ice Sheet. *Science* 258: 1921–1924.

Harlow, W.M., E.S. Harrar, and F.M. White. 1979. *Textbook on Dendrology.* McGraw-Hill, New York.

Harms, W.R. and F.T. Lloyd. 1981. Stand structure and yield relationships in a 20-yr-old loblolly pine spacing study. *Southern Journal of Applied Forestry* 5: 162–165.

Harris, W.G., K.A. Hollien, T.L. Yuan, S.R. Bates, and W.A. Acree. 1988. Nonexchangeable potassium associated with hydroxy-interlayered vermiculite from coastal plain soils. *Soil Science Society of America Journal* 52: 1486–1492.

Harrison, K., W. Post, and D.D. Richter. 1995. Recovery of soil carbon by reforestation of old-field soils. *Global Biogeochemical Cycles* 9: 449–454.

Haynes, G. 1991. *Mammoths, Mastodons, and Elephants.* Cambridge University Press, Cambridge.

Hayward, V. and E.S. Watson. 1922. *Romantic Canada.* Macmillan of Canada at St. Martin's House, Toronto.

Healy, R. 1985. *Competition for Land in the American South.* Conservation Foundation, Washington, DC.

Hedin, L.O., L. Granat, G.E. Likens *et al.* 1997. Steep declines in atmospheric base cations in regions of Europe and North America. *Nature* 367: 351–354.

Hedley, M.J., J.W.B. Stewart, and B.S. Chauhan. 1982a. Changes in the inorganic and organic soil phosphorus fractions induced by cultivation practices and by laboratory incubations. *Soil Science Society of America Journal* 46: 970–976.

Hedley, M.J., R.E. White, and R.H. Nye. 1982b. Plant-induced changes in the rhizosphere of rape (*Brassica napus* var. Emerald) seedlings. III. Changes in L value, soil phosphate fractions, and phosphatase activity. *New Phytologist* 91: 45–56.

Herbert, B.E. and P.M. Bertsch. 1995. Characterization of dissolved and colloidal organic matter in soil solution: A review. Pp. 63–88 in W. McFee and J.M. Kelly (eds.) *Carbon Forms and Functions in Forest Soils.* Soil Science Society of America, Madison, WI.

Hilgard, E.W. 1860. *Report on the Geology and Agriculture of the State of Mississippi.* State Government of Mississippi, Jackson.

Holford, I.C.R. 1981. Changes in nitrogen and organic carbon of wheat-growing soils after various periods of grazed lucerne, extended fallowing and continuous wheat. *Australian Journal of Soil Research* 31: 239–249.

Holland, H.D. 1984. *The Chemical Evolution of the Atmosphere and Oceans.* Princeton University Press, Princeton, NJ.

Holley, W.C., E. Winston, and T.J. Woofter. 1940. *The Plantation South 1934–1937.* US Government Printing Office, Washington, DC.

Homberger, E. 1995. *Historical Atlas of North America*. Penguin Books, London.

Hosner, L.R. and F.M. Hons. 1992. Reclamation of mine tailings. *Advances in Soil Science* 17: 311–350.

Houghton, R.A., E.A. Davidson, and G.M. Woodwell. 1998. Missing sinks, feedbacks, and understanding the role of terrestrial ecosystems in the global carbon balance. *Global Biogeochemical Cycles* 12: 25–34.

Hudson, C. 1976. *The Southeastern Indians*. University of North Carolina Press, Chapel Hill.

Hunt, C.B. 1986. *Surficial Deposits of the United States*. Van Nostrand Reinhold Co., New York.

Hurlbert, S. 1984. Pseudo-replication and the design of ecological field experiments. *Ecological Monographs* 54: 187–211.

Huston, M.A. 1994. *Biological Diversity: The Coexistence of Species on Changing Landscapes*. Cambridge University Press, New York.

Hutchinson, G.E. 1957. *A Treatise on Limnology*. Wiley, New York.

Jaakko Pöyry Group. 1994. *Global Fiber Resources Situation: The Challenges for the 1990s*. Unpublished report. Jaakko Pöyry Consulting, Tarrytown, NY.

Jackson, M.L. 1964. Chemical composition of soils. Pp. 71–141 in F.E. Bear (ed.) *Chemistry of the Soil*. American Chemical Society Monograph, S Series. Reinhold Publication Co., New York.

Jackson, M.L. and G.D. Sherman. 1953. Chemical weathering of minerals in soils. *Advances in Agronomy* 5: 221–317.

Jackson, W. 1980. *New Roots for Agriculture*. University of Nebraska Press, Lincoln.

Jackson, W. 1992. Living soil: *Orion* editorial. *Orion* 11(2): 3.

Jenkinson, D.S. 1971. The accumulation of nitrogen in soil left uncultivated. *Report of the Rothamsted Experiment Station 1970*, 2: 113–137.

Jenkinson, D.S. 1991. The Rothamsted classical experiments: Are they still of use? *Agronomy Journal* 83: 2–10.

Jenkinson, D.S. and J.H. Rayner. 1977. The turnover of soil organic matter in some of the Rothamsted classical experiments. *Soil Science* 123: 298–305.

Jenny, H. 1961a. Reflections on the soil acidity merry-go-round. *Soil Science Society of America Proceedings* 25: 428–432.

Jenny, H. 1961b. E. W. Hilgard and the birth of modern soil science. *Agrochimica* 3. Pisa, Italy.

Jenny, H. 1980. *The Soil Resource*. Ecological Studies 37. Springer-Verlag, New York.

Jenny, H., R.J. Arkley, and A.M. Schultz. 1969. The pygmy forest-podzol ecosystem and its dune associates of the Mendocino coast. *Madroño* 20: 60–74.

Johannessen, S. and C.A. Hastorf (eds.) 1994. *Corn and Culture in the Prehistoric New World*. Westview Press, Boulder, CO.

Johnson, C.S., E.R. Embree, and W.W. Alexander. 1935. *The Collapse of Cotton Tenancy*. University of North Carolina Press, Chapel Hill.

Johnson, D.W. and S.E. Lindberg. 1992. *Atmospheric Deposition and Nutrient Cycling in Forest Ecosystems*. Springer-Verlag, New York.

Johnson, D.W. and D.E. Todd. 1998. Effects of harvesting intensity on forest productivity and soil carbon storage. Pp. 351–363 in R. Lal (ed.) *Management of Carbon Sequestration in Soil*. CRC Press, Boca Raton, FL.

Johnson, D.W., D.W. Cole, S.P. Gessel, M.J. Singer, and R.V. Minden. 1977. Carbonic acid leaching in a tropical, temperate, subalpine, and northern forest soil. *Arctic and Alpine Research* 9: 329–343.

Johnson, D.W., J.M. Kelly, W.T. Swank *et al.* 1988. The effects of leaching and whole-tree harvesting on cation budgets of several forests. *Journal of Environmental Quality* 17: 418–424.

Johnson, D.W., M.S. Cresser, I.S. Nilsson *et al.* 1991. Soil changes in forest ecosystems: evidence for and probable causes. *Proceedings of the Royal Society of Edinburgh* 97B: 81–116.

Johnston, A.E. 1994. The Rothamsted classical experiments. Pp. 9–37 in R.A. Leigh and A.E. Johnston (eds.) *Long-term Experiments in Agricultural and Ecological Sciences*. CAB International, Wallingford, UK.

Johnston, A.E., K.W.T. Goulding, and P.R. Poulton. 1986. Soil acidification during more than 100 years under permanent grassland and woodland at Rothamsted. *Soil Use and Management* 2: 3–10.

Jones, J.B. and P.J. Mulholland. 1998. Carbon dioxide variation in a hardwood forest stream: An integrative measure of whole catchment soil respiration. *Ecosystems* 1: 183–196.

Jordan, C.F. 1998. *Working with Nature*. Harwood Academic Publishers, Amsterdam.

Jorgensen, J., C.G. Wells, and L.J. Metz. 1980. Nutrient changes in decomposing loblolly pine forest floor. *Soil Science Society of America Journal* 44: 1307–1314.

Kennedy, R.G. 1996. *Hidden Cities: The Discovery and Loss of Ancient North American Civilization*. Penguin Books, New York.

Klute, A. (ed.) 1986. *Methods of Soil Analysis, Physical and Mineralogical Methods, Part 1*. Soil Science Society of America, Madison, WI.

Knoepp, J.D. and W.T. Swank. 1994. Long-term soil chemical changes in aggrading forest ecosystems. *Soil Science Society of America Journal* 58: 425–331.

Körschens, M. (ed.) 1994. *Der Statische Duengungsversuch Bad Lauchstadt nach 90 Jahren*. B.G. Teubner Verlagsgesellschaft, Stuttgart-Leipzig.

Kuhn, T.S. 1970. *The Structure of Scientific Revolutions*. University of Chicago Press, IL.

Kurten, B. and E. Anderson. 1980. *Pleistocene Mammals of North America*. Columbia University Press, New York.

Låg, J. 1968. Relationships between the chemical composition of the precipitation and the contents of exchangeable ions in the humus layer of natural soils. *Acta Agriculturae Scandinavica* 18: 148–152.

Lal, R. and B.A. Stewart (eds.) 1995. *Soil Management: Experimental Basis for Sustainability and Environmental Quality*. CRC Press, Boca Raton, FL.

Lapeyrie, F., G.A. Chilvers, and C.A. Bhem. 1987. Oxalic acid synthesis by the mycorrhizal fungus *Paxillus involutus* (Battsch. ex. Fr.) Fr. *New Phytologist* 106: 139–146.

Larson, L.H. 1980. *Aboriginal Subsistence Technology on the Southeastern Coastal Plain during the late Prehistoric Period*. University Presses of Florida, Gainesville.

Lebedeva, E.V., N.N. Lyalikova, and Y.Y. Bugel'skii. 1979. Participation of nitrifying bacteria in the weathering of serpentized ultrabasic rock. *Mikrobiologia* 47: 898–904.

Lefler, H.T. (ed.) 1967. *A New Voyage to Carolina: John Lawson*. University of North Carolina Press, Chapel Hill.

Leigh, R.A. and A.E. Johnston (eds.) 1994. *Long-term Experiments in Agricultural and Ecological Sciences*. CAB International, Wallingford, UK.

Leopold, A. 1949. *Sand County Almanac*. Oxford University Press, Oxford.

Lepsch, I.F., S.W. Buol, and R.B. Daniels. 1977. Soil-landscape relationships in the Occidental Plateau of Sao Paulo State, Brazil: I. Geomorphic surfaces and soil mapping units. *Soil Science Society of America Journal* 41: 104–109.

Leyval, C. and J. Berthelin. 1991. Weathering of a mica by roots and rhyzospheric microorganisms of pine. *Soil Science Society of America Journal* 55: 1009–1016.

Likens, G.E. (ed.) 1989. *Long-term Studies in Ecology: Approaches and Alternatives.* Springer-Verlag, New York.

Likens, G.E., F.H. Bormann, R.S. Pierce, J.S. Eaton, and N.M. Johnson. 1977. *Biogeochemistry of a Forested Ecosystem.* Springer-Verlag, New York.

Likens, G.E., C.T. Driscoll, and D.C. Buso. 1992. Long-term effects of acid rain: Response and recovery of a forest ecosystem. *Science* 272: 244–246.

Lindberg, S.E., G.M. Lovette, D.D. Richter, and D.W. Johnson. 1986. Atmospheric deposition and canopy interactions of major ions in a forest. *Science* 231: 141–145.

Lutz, H.J. and R.F. Chandler. 1946. *Forest Soils.* Wiley, New York.

Mabry, M.L. 1981. *Union County Heritage.* Winston-Salem, NC.

McCollum, R.E. 1991. Buildup and decline of soil phosphorus: 30 year trends on a Typic Umbraquult. *Agronomy Journal* 83: 77–85.

McCracken, R.J., R.B. Daniels, and W.E. Fulcher. 1989. Undisturbed soils, landscapes, and vegetation in a North Carolina Piedmont virgin forest. *Soil Science Society of America Journal* 53: 1146–1152.

McGill, W.B., K.R. Cannon, J.A. Robertson, and F.D. Cook. 1986. Dynamics of soil microbial biomass and water-soluble organic C in Breton after 50 years of cropping to two rotations. *Canadian Journal of Soil Science* 66: 1–19.

Magnuson, J.J. 1990. The invisible present. *BioScience* 40: 495–501.

Marion, G.M. 1979. Biomass and nutrient removal in long-rotation stands. Pp. 98–110 in A. Leaf (ed.) *Impact of Intensive Harvesting on Forest Nutrient Cycling.* State University of New York at Syracuse.

Markewitz, D. 1996. *Soil Acidification, Soil Potassium Availability, and Biogeochemistry of Aluminum and Silicon in a 34-year old Loblolly Pine* (Pinus taeda L.) *Ecosystem in the Calhoun Experimental Forest, South Carolina.* PhD Dissertation, Duke University, Durham, NC.

Markewitz, D. and D.D. Richter. 2000. Long-term potassium availability from a Kanhapludult to an aggrading loblolly pine ecosystem. *Forest Ecology and Management* 130: 109–129.

Markewitz, D., D.D. Richter, H.L. Allen, and J.B. Urrego. 1998. Three decades of observed soil acidification at the Calhoun Experimental Forest: Has acid rain made a difference? *Soil Science Society of America Journal* 62: 1428–1439.

Marks, P.L. and F.H. Bormann. 1972. Revegetation following forest cutting: Mechanisms for return to steady-state nutrient cycling. *Science* 176: 914–915.

Marsh, G.P. 1864. *Man and Nature; or, Physical Geography as Modified by Human Action.* Charles Scribner, New York.

Mayer, J.P. (ed.) 1960. *Journey to America by A. de Tocqueville.* Faber and Faber, London.

Meade, R.H. and S.W. Trimble. 1974. Changes in sediment loads of the Atlantic drainage of the United States. *International Association of Hydrological Sciences* 113: 99–104.

Mehlich, A. 1978. New extractant for soil test evaluation of phosphorus, potassium, magnesium, calcium, sodium, manganese, and zinc. *Communications in Soil Science and Plant Analysis* 9: 477–492.

Melillo, J.M., J.D. Aber, P.A. Steudler, and J.P. Schimel. 1983. Denitrification potentials in a successional sequence of northern hardwood forest stands. Pp. 217–228 in R. Hallberg (ed.) *Environmental Biogeochemistry*. Swedish Natural Science Research Council, Stockholm.

Melillo, J.M., C.A. Palm, R.A. Houghton, G.M. Woodwell, and N. Myers. 1985. A comparison of two recent estimates of disturbance in tropical forests. *Environmental Conservation* 12: 37–40.

Miller, H.G., J.M. Cooper, J.D. Miller, and O.J.L. Pauline. 1978. Nutrient cycles in pine and their adaptation to poor soils. *Canadian Journal of Forest Research* 9: 19–26.

Milne, G. 1935. Composite units for the mapping of complex soil associations. *Transactions of the 3rd International Congress of Soil Science Oxford* 1: 345–347.

Mitchell, C.C. 1988. *New Information from Old Rotation*. Highlights in Agricultural Research 35. Alabama Agricultural Experiment Station, Auburn University, AL.

Mitchell, C.C., R.L. Westerman, J.R. Brown, and T.R. Peck. 1991. Overview of long-term agronomic research. *Agronomy Journal* 83: 24–29.

Mitchell, C.C., F.J. Arriaga, J.A. Entry *et al.* 1996. *The Old Rotation*. Auburn University Agricultural Experiment Station, AL.

Muller, J. 1978. The Kincaid system: Mississippian settlement in the environs of a large site. Pp. 269–292 in B.D. Smith (ed.) *Mississippian Settlement Patterns*. Academic Press, New York.

Murphy, J. and J.P. Riley. 1962. A modified single solution method for the determination of phosphate in natural waters. *Analytica Chimica Acta* 27: 31–36.

Nambiar, E.K.S. 1996. Sustained productivity of forests is a continuing challenge to soil science. *Soil Science Society of America Journal* 60: 1629–1642.

Nepsted, D., C.R. de Carvalho, E.A. Davidson *et al.* 1994. The role of deep roots in the hydrological and carbon cycles of Amazonian forests. *Nature* 372: 666–669.

Newman, E.I. 1997. Phosphorus balance of contrasting farming systems: Past and present. Can food production be sustainable? *Journal of Applied Ecology* 34: 1334–1347.

Novais, R. and E.J. Kamprath. 1978. Phosphorus supplying capacities of previously heavily fertilized soils. *Soil Science Society of America Journal* 42: 931–935.

Nowak, C.A., R.B. Downard, and E.H. White. 1991. Potassium trends in red pine plantations at Pack Forest, New York. *Soil Science Society of America Journal* 55: 847–850.

Nye, P.H. and D.J. Greenland. 1960. *The Soil under Shifting Cultivation*. Communication 51, Commonwealth Bureau of Soils, London.

Oades, J.M. 1994. The retention of organic matter in soils. *Biogeochemistry* 5: 35–70.

Olsen, S.R., C.V. Cole, F.S. Watanabe, and L.A. Dean. 1954. *Estimation of Available Phosphorus in Soils by Extraction with Sodium Bicarbonate*. United States Department of Agriculture Circular 939. United States Government Printing Office, Washington, DC.

O'Neill, J.S. 1985. Cenozoic fluctuations in biotic parts of the global carbon cycle. Pp. 377–396 in E.T. Sundquist and W.S. Broeker (eds.) *The Carbon Cycle and Atmospheric CO_2: Natural Variations Archean to Present*. Geophysical Monographs 32. American Geophysical Union, Washington, DC.

O'Neill, K.P. 2000. *Changes in Carbon Dynamics following Wildfires from Forest*

Soils in the Interior of Alaska. PhD Dissertation, Duke University, Durham, NC.

O'Neill, R.V., D.L. DeAngelis, J.B. Waide, and T.F.H. Allen 1986. *A Hierarchical Theory of Ecosystems.* Princeton University Press, Princeton, NJ.

Oosting, H.J. 1942. An ecological analysis of the plant communities of Piedmont, North Carolina. *American Midland Naturalist* 28: 1–126.

Ovington, J.D. 1962. Quantitative ecology and the woodland ecosystem concept. *Advances in Ecological Research* 1: 103–192.

Owens, L.L. 1971. *Saints of Clay: The Shaping of South Carolina Baptists.* R.L. Bryan Co., Columbia, SC.

Page, A.L. (ed.) 1982. *Methods of Soil Analysis, Chemical and Microbiological Properties, Part 2.* Soil Science Society of America, Madison, WI.

Pastor, J. and W.M. Post. 1986. Influence of climate, soil moisture, and succession on forest carbon and nitrogen cycles. *Biogeochemistry* 2: 3–27.

Paton, T.R., P.B. Humpheries, and G.S. Mitchell. 1995. *Soils.* Yale University Press, New Haven, CT.

Patterson, C.C. and T.W. Cooney. 1986. Sediment transport and deposition in Lakes Marion and Moultrie, South Carolina. Pp. 1336–1351 in *Third International Symposium on River Sedimentation.* University of Mississippi, Oxford, USA.

Pavich, M.J. 1985. Appalachian piedmont morphogenesis: Weathering, erosion, and Cenozoic uplift. Pp. 299–319 in M. Morisawa and J.T. Hack (eds.) *Tectonic Geomorphology.* Allen and Unwin, Boston, MA.

Pavich, M.J. 1986. Processes and rates of saprolite production and erosion on a foliated granite rock of the Virginia Piedmont. Pp. 552–590 in S.M. Colman and D.P. Dethier (eds.) *Rates of Chemical Weathering of Rocks and Minerals.* Academic Press, Orlando, FL.

Peebles, C.S. 1978. Determinants of settlement size and location in the Moundville phase. Pp. 369–416 in B.D. Smith (ed.) *Mississippian Settlement Patterns.* Academic Press, New York.

Peters, R.H. 1991. *A Critique for Ecology.* Cambridge University Press, Cambridge.

Peterson, W.H. and G.H. Aull. 1945. *A Pattern of Agricultural Production in South Carolina after the War.* Pamphlet. Clemson University Agricultural Experiment Station, SC.

Pickett, S.T.A. 1989. Space-for-time substitution as an alternative to long-term studies. Pp. 110–135 in G.E. Likens (ed.) *Long-term Studies in Ecology: Approaches and Alternatives.* Springer-Verlag, New York.

Pickett, S.T.A. 1991. Long-term studies: Past experience and recommendations for the future. Pp. 71–88 in P.G. Risser (ed.) *Long-term Ecological Research.* Scientific Committee on Problems of the Environment (SCOPE) of the International Council of Scientific Unions (IUCN). Wiley, New York.

Pierre, W.H., J. Meisinger, and J.R. Birches. 1970. Cation–anion balance in crops as a factor in determining the effects of nitrogen fertilizers on soil acidity. *Agronomy Journal* 62: 106–112.

Portella, E.A. 1993. Potassium supplying capacity of northeastern Portuguese soils. Pp. 57–64 in M.A.C. Trafoso and M.L. van Beusichem (eds.) *Optimization of Plant Nutrients.* Kluwer Academic Publishers, Dordrecht, The Netherlands.

Post, W.M. and L.K. Mann. 1990. Changes in soil organic carbon and nitrogen as a result of cultivation. Pp. 401–406 in A.F. Bouman (ed.) *Soil and the Greenhouse Effect.* Wiley, London.

Post, W.M., J. Pastor, A.W. King, and W.R. Emanuel. 1992. Aspects of the interaction between vegetation and soil under global change. *Water, Air, and Soil Pollution* 64: 345–363.

Post, W.M., A.W. King, and S.D. Wullshleger. 1997. Historic variations in terrestrial biospheric carbon storage. *Global Biogeochemical Cycles* 11: 99–109.

Powell, D.S., J.L. Faulkner, D.R. Darr, Z. Zhu, and D.W. MacCleery. 1993. *Forest Resources of the United States, 1992.* USDA Forest Service General Technical Report RM-234. Rocky Mountain Forest and Range Experiment Station, Fort Collins, CO.

Powers, R. and K. Van Cleve. 1991. Long-term ecological research in temperate and boreal forest ecosystems. *Agronomy Journal* 83: 11–24.

Pregitzer, K.S. and B.J. Palik. 1997. Changes in ecosystem carbon 46 years after establishing red pine (*Pinus resinosa* Ait.) on abandoned agricultural land in the Great Lakes Region. Pp. 263–270 in E.A. Paul, K. Paustian, E.T. Elliot, and C.V. Cole (eds.) *Soil Organic Matter in Temperate Agroecosystems.* CRC Press, New York.

Puckett, L.J. 1994. *Non-point and Point Sources of Nitrogen in Major Watersheds of the United States.* US Geological Survey Water Resources Investigations Report 94-4001. Washington, DC.

Qualls, R.G. and B.L. Haines. 1991. Geochemistry of dissolved organic nutrients in water percolating through a forest ecosystem. *Soil Science Society of America Journal* 55: 1112–1123.

Radford, A.E., H.E. Ahles, and C.R. Bell. 1968. *Manual of the Vascular Flora of the Carolinas.* University of North Carolina Press, Chapel Hill.

Rennie, P.J. 1955. Uptake of nutrients by mature forest trees. *Plant and Soil* 7: 49–95.

Retallack, G.J. 1990. *Soils of the Past.* Unwin Hyman, London.

Retallack, G.J. 1992. Paleozoic paleosols. Pp. 543–564 in I.P. Martini and W. Chesworth (eds.) *Weathering, Soils, and Paleosols.* Elsevier, New York.

Reuss, J.O. and D.W. Johnson. 1986. *Acid Deposition and the Acidification of Soils.* Springer-Verlag, New York.

Richards, B.N. 1973. Nitrogen fixation in the rhizospheres of conifers. *Soil Biology and Biochemistry* 5: 149–152.

Richter, D.D. 1986. Sources of acidity in some forested Udults. *Soil Science Society of America Journal* 50: 1584–1589.

Richter, D.D. and L.I. Babbar. 1991. Soil diversity in the tropics. *Advances in Ecological Research* 21: 316–389.

Richter, D.D., and D. Markewitz. 1995a. Atmospheric deposition and soil resources of the southern pine forest. Pp. 315–336 in S. Fox and R. Mikler (eds.) *Impact of Air Pollutants on Southern Pine Forests.* Springer-Verlag, New York.

Richter, D.D. and D. Markewitz. 1995b. How deep is soil? *BioScience* 45: 600–609.

Richter, D.D. and D. Markewitz. 1996. Soil carbon dynamics during the growth of old-field loblolly pine at the Calhoun Experimental Forest, USA. Pp. 397–407 in D. Powlson, P. Smith, and I. Smith (eds.) *Evaluation of Soil Organic Matter Models.* NATO Advanced Science Series. Springer-Verlag, Berlin.

Richter, D.D., C.W. Ralston, and W.R. Harms. 1982. Prescribed fire: Effects on water quality and forest nutrient cycling. *Science* 215: 661–663.

Richter, D.D., D. Markewitz, C.G. Wells *et al.* 1994. Soil chemical change during three decades in an old-field loblolly pine (*Pinus taeda* L.) ecosystem. *Ecology* 75: 1463–1473.

Richter, D.D., K. Korfmacher, and R. Nau. 1995a. *Decreases in Yadkin River Basin Sedimentation: Statistical and Geographic Time-trend Analyses, 1951 to 1990*. North Carolina Water Resources Research Institute Report 297. Raleigh, NC.

Richter, D.D., D. Markewitz, C.G. Wells *et al.* 1995b. Carbon cycling in an old-field pine forest: Implications for the missing carbon sink and for the concept of soil. Pp. 233–251 in W. McFee and J.M. Kelly (eds.) *Carbon Forms and Functions in Forest Soils*. Soil Science Society of America, Madison, WI.

Richter, D.D., D. Markewitz, S.A. Trumbore, and C.G. Wells. 1999. Rapid accumulation and turnover of soil carbon in a re-establishing forest. *Nature* 400: 56–58.

Richter, D.D., K.P. O'Neill, and E. Kaschiske. 2000a. Stimulation of decomposition following wildfires in boreal black spruce (*Picea mariana* L.) ecosystems: A hypothesis. Pp. 197–213 in E. Kaschiske and B. Stocks (eds.) *Fire, Climate Change, and Carbon Cycling in the Boreal Forest*. Springer-Verlag, New York.

Richter, D.D., D. Markewitz, P.R. Heine *et al.* 2000b. Legacies of agriculture and forest regrowth in the nitrogen of old-field soils. *Forest Ecology and Management* 138: 233–248.

Risser, P. (ed.) 1991. *Long-term Ecological Research: An International Perspective*. Scientific Committee on Problems of the Environment (SCOPE) of the International Council of Scientific Unions (IUCN). Wiley, New York.

Robertson, G.P., C.S. Bledsoe, D.C. Coleman, and P. Sollins (eds.) 1999. *Standard Soil Methods for Long-term Ecological Research*. Oxford University Press, Oxford.

Rodale, J.I. 1945. *Pay Dirt, Farming and Gardening with Composts*. Devin-Adair Co., New York.

Ruark, G.A. 1993. Modeling soil temperature effects on in situ decomposition rates for fine roots of loblolly pine. *Forest Science* 39: 118–129.

Ruffin, E. 1852. *An Essay on Calcareous Manures*. J.W. Randolph, Richmond, VA.

Russell, J.S. 1960. Soil fertility changes in the long-term experimental plots at Kybybolite, South Australia. I. Changes in pH, total nitrogen, organic carbon, and bulk density. *Australian Journal of Agricultural Research* 11: 902–926.

Rust, R.H. 1983. Alfisols. Pp. 253–281 in L.P. Wild, N.W. Smeck, and G.F. Hall (eds.) *Pedogenesis and Soil Taxonomy. Vol. II: The Soil Orders*. Elsevier, Amsterdam, The Netherlands.

Ruxton, B.P. and L. Berry. 1957. Weathering of granite and associated erosional features in Hong Kong. *Geological Society of America Bulletin* 8: 1263–1292.

Ryan, M.G., D. Binkley, and H.H. Fownes. 1997. Age-related decline in forest productivity: Pattern and process. *Advances in Ecological Research* 27: 213–262.

Sánchez, P.A. 1976. *Properties and Management of Soils in the Tropics*. Wiley, New York.

Sánchez, P.A. 1994. Tropical soil fertility research: Towards a second paradigm. Pp. 65–88 in *Transactions of 15th World Congress of Soil Science*, Vol. 1. Acapulco, Mexico.

Sánchez, P.A., C.A. Palm, L.T. Szott, and C.B. Davey. 1985. Trees as soil improvers in the humid tropics? Pp. 331–362 in M.G.C. Cannell (ed.) *Attributes of Trees as Crop Plants*. Institute of Terrestrial Ecology, Midlothian, Scotland, UK.

Scarry, C.M. (ed.) 1993. *Foraging and Farming in the Eastern Woodlands*. University Press of Florida, Gainesville.

Schiffman, P.M. and W. Johnson. 1991. Phytomass and detrital carbon storage during forest regrowth in southern United States piedmont. *Canadian Journal of Forest Research* 19: 69–78.

Schjonning, P., B.T. Christensen, and B. Carstensen. 1994. Physical and chemical properties of a sandy loam receiving animal manure, mineral fertilizer or no fertilizer for 90 years. *European Journal of Soil Science* 45: 257–268.

Schimel, D.S. 1995. Terrestrial ecosystems and the carbon cycle. *Global Change Biology* 1: 77–91.

Schimel, D.S., J. Mellilo, H.Q. Tian *et al.* 2000. Contributions of increasing CO_2 and climate to carbon storage by ecosystems in the United States. *Science* 287: 2004–2006.

Schlesinger, W.H. 1990. Evidence from chronosequence studies for a low carbon-storage potential of soils. *Nature* 348: 232–234.

Schlesinger, W.H. 1997. *Biogeochemistry*. 2nd edn. Academic Press, New York.

Schmidt, J.P., S.W. Buol, and E.J. Kamprath. 1996. Soil phosphorus dynamics during seventeen years of continuous cultivation: Fractionation analyses. *Soil Science Society of America Journal* 60: 1168–1172.

Schmidt, J.P., S.W. Buol, and E.J. Kamprath. 1997. Soil phosphorus dynamics during 17 years of continuous cultivation: A method to estimate long-term P availability. *Geoderma* 78: 59–70.

Schoeneberger, P. 1995. Soils, geomorphology, and land use in the southeastern United States. Pp. 58–82. in S. Fox and R.A. Mikler (eds.) *Impact of Air Pollutants on Southern Pine Forests*. Ecological Studies 118. Springer-Verlag, New York.

Schoeninger, M.J. and M.R. Schurr. 1994. Interpreting carbon stable isotope ratios. Pp. 55–66 in S. Johannessen and C.A. Hastorf (eds.) *Corn and Culture in the Prehistoric New World*. Westview Press, Boulder, CO.

Schwertmann, U. and A.J. Herbillon. 1992. Some aspects of fertility associated with the mineralogy of highly weathered tropical soils. pp. 47–59 in R. Lal and P.A. Sanchez (eds.) *Myths and Science of Soils in the Tropics*. SSSA Special Publication 29. Soil Science Society of America, Madison, WI.

Shelton, J.E. and N.T. Coleman. 1968. Inorganic phosphorus fractions and their relationship to residual value of large applications of phosphorus on high phosphorus fixing soils. *Soil Science Society of America Proceedings* 32: 91–94.

Sheridan, R.C. 1979. Chemical fertilizers in southern agriculture. *Agricultural History* 53: 308–318.

Shugart, H.H. 1984. *A Theory of Forest Dynamics: The Ecological Implications of Forest Succession Models*. Springer-Verlag, New York.

Sibirtzev, N.M. 1914. *Soil Science*. (Transl. N. Kaner.) Israel Program for Scientific Translations, Jerusalem and United States Department of Agriculture, Washington, DC.

Simonson, R.M. 1959. Outline of a generalized theory of soil genesis. *Soil Science Society of America Proceedings* 23: 152–156.

Smith, B.D. 1985. Mississippian patterns of subsistence and settlement. Pp. 64–79 in R.R. Badger and L.A. Clayton (eds.) *Alabama and the Borderlands*. The University of Alabama Press, University.

Smith, B.D. 1989. Origins of agriculture in eastern North America. *Science* 246: 1566–1571.

Smith, P., D.S. Powlson, J.U. Smith, and E.T. Elliott (eds.) 1997. Evaluation and comparison of soil organic matter models. *Geoderma* 81: 1–225.

Smyth, T.J. and D.K. Cassel. 1995. Synthesis of long-term soil management research on Ultisols and Oxisols in the Amazon. Pp. 13–60 in R. Lal and B.A. Stewart (eds.) *Soil Management: Experimental Basis for Sustainability and Environmental Quality*. CRC Lewis Pubishers, Boca Raton, FL.

Soil Survey Staff. 1975. *Soil Taxonomy*. Soil Conservation Service, United States Department of Agriculture, Washington, DC.

Soil Survey Staff. 1992. *Keys to Soil Taxonomy*. Cornell University, Ithaca, NY.

Soil Survey Staff. 1998. *Keys to Soil Taxonomy*. Natural Resources Conservation Service, Washington, DC.

Sollins, P., C.C. Grier, F.M. McCorison, K. Cromack, R. Fogel, and R.L. Fredriksen. 1980. The internal element cycles of an old-growth Douglas-fir ecosystem in western Oregon. *Ecological Monographs* 50: 261–285.

Sollins, P., G. Spycher, and C. Topik. 1983. Processes of soil organic-matter accretion at a mudflow chronosequence, Mt. Shasta, California. *Ecology* 64: 1273–1282.

Sparks, C.E. 1967. *A History of Padgett's Creek Baptist Church*. Counts Printing Company, Union, SC.

Sparks, D.L. 1987. Potassium dynamics in soils. *Advances in Soil Science* 6: 1–63.

Spears V.S. 1974. *The Fairforest Story: History of the Fairforest Baptist Church and Community*. Crabtree Press, Inc., Charlotte, NC.

Stark, J. and S.C. Hart. 1998. High rates of nitrification and nitrate turnover in undisturbed coniferous forests. *Nature* 385: 61–64.

Stevenson, G. 1959. Fixation of nitrogen by non-nodulated seed plants. *Annals of Botany*, n.s. 23: 622–635.

Stolt, M.H., J.C. Baker, and T.W. Simpson. 1992. Characterization and genesis of saprolite derived from gneissic rocks of Virginia. *Soil Science Society of America Journal* 56: 531–539.

Stone, E.L. 1975. Effects of species on nutrient cycles and soil change. *Philosophical Transactions of the Royal Society of London* B271: 149–162.

Stone, E.L. 1979. Nutrient removals by intensive harvest – Some research gaps and opportunities. Pp. 366–386 in A. Leaf (ed.) *Impact of Intensive Harvesting on Forest Nutrient Cycling*. State University of New York at Syracuse.

Stone, E.L. and R. Kszystyniak. 1977. Conservation of potassium in the *Pinus resinosa* ecosystem. *Science* 198: 192–194.

Strain, B.R. 1985. Physiological and ecological controls on carbon sequestering in terrestrial ecosystems. *Biogeochemistry* 1: 219–232.

Stuanes, A., H. Van Miegroet, D.W. Cole, and G. Abrahamson. 1992. Recovery from acidification. Pp. 467–494 in D.W. Johnson and S.E. Lindberg (eds.) *Atmospheric Deposition and Nutrient Cycling in Forest Ecosystems*. Springer-Verlag, New York.

Stumm, W. and J.J. Morgan. 1981. *Aquatic Chemistry*. Wiley, New York.

Sundquist, E.T. 1993. The global carbon dioxide budget. *Science* 259: 934–941.

Swank, W.T. and D.A. Crossley (eds.) 1988. *Forest Hydrology and Ecology at Coweeta*. Springer-Verlag, New York.

Swanton, J.R. 1946. *The Indians of the Southeastern United States*. Bulletin 137. Smithsonian Institution Bureau of American Ethnology, Washington, DC.

Swift, J. 1753 (1976). *Gulliver's Travels*. Easton Press, Norwalk, CT.

Switzer, G.E., and L.E. Nelson. 1972. Nutrient accumulation and cycling in loblolly pine (*Pinus taeda* L.) plantation ecosystems: The first twenty years. *Soil Science Society of America Proceedings* 36: 143–147.

Tamm, C.O. and L. Hallbacken. 1986. Changes in soil pH over a 50-yr period under different forest canopies in SW Sweden. *Water, Air, and Soil Pollution* 31: 337–341.

Taylor, J. 1813. *The Arator; Being a Series of Agricultural Essays, Practical and Political.* Georgetown, DC.

Tennessee Valley Authority. 1962. *Reforestation and Erosion Control Influences upon the Hydrology of the Pine Tree Branch Watershed (1942–1960).* Washington, DC.

Thomas, G.W., and W.L. Hargrove. 1984. The chemistry of soil acidity. Pp. 3–49 in F. Adams (ed.) *Soil Acidity and Liming.* 2nd edn. Soil Science Society of America, Madison, WI.

Thomas, G.W. and D.E. Peaslee. 1973. Testing soils for phosphorus. Pp. 115–132 in. L.M. Walsh and J.D. Beaton (eds.) *Soil Testing and Plant Analysis.* Soil Science Society of America, Madison, WI.

Thorne, J.F. and S.P. Hamburg. 1985. Nitrification potentials of an old-field chronosequence in Compton, New Hampshire. *Ecology* 66: 1333–1338.

Thornton, R., J. Warren, and T. Miller. 1992. Depopulation in the southeast after 1492. Pp. 187–196 in J.W. Verano and D.H. Uberlaker (eds.) *Disease and Demography in the Americas.* Smithsonian Institution Press, Washington, DC.

Thurston, J., E.D. Williams, and A.E. Johnston. 1976. Modern developments in an experiment on permanent grassland started in 1856: Effects of fertilizers and lime on botanical composition and crop and soil analysis. *Annales Agronomiques* 27: 1043–1082.

Tiessen, H. and J.O. Moir. 1993. Characterization of available P by sequential extraction. Pp. 75–86 in M.R. Carter (ed.) *Soil Sampling and Analysis.* Canadian Society of Soil Science/Lewis Publishers, Boca Raton, FL.

Tiessen, H., J.W.B. Stewart, and J.O. Moir. 1983. Changes in organic and inorganic phosphorus composition of two grassland soils and their particle size fractions during 60–70 years of cultivation. *Journal of Soil Science* 34: 815–823.

Tiessen, H., I.H. Salceda, and E.V.S.B. Sampaio. 1992. Nutrient and soil organic matter dynamics under shifting cultivation in semi-arid northeastern Brazil. *Agriculture, Ecosystems, and Environment* 38: 139–151.

Tilman, D., M.E. Dodd, J. Silvertown, P.R. Poulton, A.E. Johnston, and M.J. Crawley. 1994. The Park Grass experiment: Insights from the most long-term ecological study. Pp. 287–303 in R.A. Leigh and A.E. Johnston (eds.) *Long-term Experiments in Agricultural and Ecological Sciences.* CAB International, Wallingford, UK.

Tinker, P.B. 1994. Monitoring environmental change through networks. Pp. 407–421 in R.A. Leigh and A.E. Johnston (eds.) *Long-term Experiments in Agricultural and Ecological Sciences.* CAB International, Wallingford, UK.

Tisdale, S.L., W.L. Nelson, and J.D. Beaton. 1985. *Soil Fertility and Fertilizers.* Macmillan, London.

Tocqueville, A. De. 1838. *Democracy in America.* 3rd edn. (Transl. H. Reeves.) Saunders and Otley, London.

Torn, M.S., S.E. Trumbore, O.A. Chadwick, P.M. Vitousek, and D.M. Hendricks. 1997. Mineral control of soil carbon storage and turnover. *Nature* 389: 170–173.

Trimble, S.W. 1974. *Man-induced Soil Erosion on the Southern Piedmont, 1700–1970.* Soil Conservation Society, Ankeny, IA.

Troedsson, T. 1980. Long-term changes of forest soils. *Annales Agriculturae Fenniae* 19: 81–84.

Trumbore, S.A. 1996. Measurement of cosmogenic isotopes by accelerator mass spectrometry: applications to soil science. Pp. 311–340 in T. Boutton and S. Yamasaki (eds.) *Mass Spectrometry of Soils*. Dekker, New York.

Tull, J. 1731. *The Horse Hoeing Husbandry*. J.M. Cobbett, London.

Ugolini, F.C. and R.S. Sletten. 1991. The role of proton donors in pedogenesis as revealed by soil solution studies. *Soil Science* 151: 59–75.

Ulrich, B., R. Mayer, and P.K. Khanna. 1980. Chemical changes due to acid precipitation in a loess-derived soil in central Europe. *Soil Science* 130: 193–199.

UN-FAO. 1998. *FAO Statistical Databases*. http://apps.fao.org/.

Urrego, M.J.B. 1993. *Nutrient Accumulation in Biomass and Forest Floor of a 34-year-old Loblolly Pine Plantation*. MS Thesis, North Carolina State University, Raleigh, NC.

van Breemen, N. and P. Buurman. 1998. *Soil Formation*. Kluwer Academic Publishers, Dordrecht, The Netherlands.

van Breemen, N., P.A. Burrough, E.J. Velthorst *et al.* 1982. Acidification from atmospheric ammonium sulphate in forest canopy throughfall. *Nature* 299: 548–550.

van Donk, J. 1976. O^{18} record of the Atlantic Ocean for the entire Pleistocene Epoch. Pp. 147–163 in R.M. Cline and J.D. Hays (eds.) *CLIMAP: Investigation of Late Quaternary Paleoceanography and Paleoclimatology*. Memoir 145. Geological Society of America, Boulder, CO.

van Lear, D., P.R. Kapeluck, and M.M. Parker. 1995. Distribution of carbon in a Piedmont soil as affected by loblolly pine management. Pp. 489–501 in W. McFee and J.M. Kelly (eds.) *Carbon Forms and Functions in Forest Soils*. Soil Science Society of America, Madison, WI.

van Miegroet, H. and D.W. Cole. 1985. Acidification sources in red alder and Douglas-fir soils: Importance of nitrification. *Soil Science Society of America Journal* 49: 1274–1279.

Vance, E.D. 1998. *Agricultural Site Productivity: Principles Derived from Long-term Experiments and their Implications for Managed Forests*. Technical Bulletin 766. National Council for Air and Stream Improvement, Research Triangle Park, NC.

Vance, R.B. 1929. *Human Factors in Cotton Culture*. University of North Carolina, Chapel Hill.

Vanotti, M.B. and L.G. Bundy. 1995. Soil organic matter dynamics in the North American corn belt: The Arlington plots. Pp. 409–418 in D. Powlson, P. Smith, and I. Smith (eds.) *Evaluation of Soil Organic Matter Models*. NATO Advanced Science Series. Springer-Verlag, Berlin.

Vitousek, P.M. and S.W. Andariese. 1986. Microbial transformations of labelled nitrogen in a clear-cut forest. *Oecologia* 68: 601–605.

Vitousek, P.M. and H. Farrington. 1997. Nutrient limitation and soil development: Experimental test of a biogeochemical theory. *Biogeochemistry* 37: 63–75.

Vitousek, P.M. and W.A. Reiners. 1975. Ecosystem succession and nutrient retention: A hypothesis. *BioScience* 25: 376–381.

Vitousek, P.M., D.R. Turner, and K. Kitayama. 1995. Foliar nutrients during long-term soil development in Hawaiian montane rain forest. *Ecology* 76: 712–720.

Vogt, K.A., C.C. Grier, C.E. Meier, and M.R. Keys. 1983. Organic matter and nutrient dynamics in forest floors in young and mature *Abies amabilis* stands in western Washington, as affected by fine-root input. *Ecological*

Monographs 53: 139–157.

von Liebig, J. 1843. *Chemistry in its Application to Agriculture and Physiology.* J.M. Campbell, Philadelphia, PA.

Walker, L.R., J.C. Zasada, and F.S. Chapin. 1986. The role of life history processes in primary succession on an Alaskan floodplain. *Ecology* 67: 1243–1253.

Walker, T.W. and J.K. Syers. 1976. The fate of phosphorus during pedogenesis. *Geoderma* 15: 1–19.

Wallace, H.A. and W.L. Brown. 1988. *Corn and its Early Fathers.* Iowa State University Press, Ames.

Way, J.T. 1850. On the power of soils to absorb manure. *Journal of the Royal Agricultural Society of England* 11: 313–379.

Webb, B.B., B.B. Tucker, and R.L. Westerman. 1980. *The Magruder Plots: Taming the Prairie through Research.* Oklahoma Agricultural Experiment Station Bulletin B-750. Stillwater, OK.

Wells, C.G. 1971. Effects of prescribed burning on soil chemical properties and nutrient availability. Pp. 86–97 in *Prescribed Burning Symposium.* USDA Forest Service, Southeast Forest Experiment Station, Asheville, NC.

Wells, C.G. and J.R. Jorgensen. 1975. Nutrient cycling in loblolly pine plantations. Pp. 137–158 in B. Bernier and C.H. Winget (eds.) *Forest Soils and Forest Land Management* (Fourth North American forest soils conference). Laval University Press, Quebec, Canada.

Wells, C.G. and J.R. Jorgensen. 1979. Effect of intensive harvesting on nutrient supply and sustained productivity. Pp. 212–230 in A. Leaf (ed.) *Impact of Intensive Harvesting on Forest Nutrient Cycling.* State University of New York at Syracuse.

Wells, C.G., D.M. Crutchfield, N.M. Berenyi, and C.B. Davey. 1973. *Soil and Plant Guidelines for Phosphorus Fertilization of Loblolly Pine.* Research Paper SE-110. USDA Forest Service, Asheville, NC.

Westerman, R.L. (ed.) 1990. *Soil Testing and Plant Analysis.* 3rd edn. Soil Science Society of America, Madison, WI.

Whitney, G.G. 1994. *From Coastal Wilderness to Fruited Plain.* Cambridge University Press, Cambridge.

Wilde, S.A. 1958. *Forest Soils.* Ronald Press, New York.

Williams, M. 1989. *Americans and their Forests.* Cambridge University Press, Cambridge.

Winogradsky, S. 1938. La microbiologie oecologique. *Annales de l'Institut Pasteur* 61: 731–755.

Wolfe, J.A. 1985. Distribution of major vegetational types during the Tertiary. Pp. 357–375 in E.T. Sundquist and W.S. Broeker (eds.) *The Carbon Cycle and Atmospheric CO_2: Natural Variations Archean to Present.* Geophysical Monographs 32. American Geophysical Union, Washington, DC.

Wood, P.H. 1989. The changing population of the colonial south: An overview by race and region, 1685–1790. Pp. 35–103 in P.H. Wood, G.A. Waselkov, and M.T. Hatley (eds.) *Pohatin's Mantle: Indians in the Colonial Southeast.* University of Nebraska Press, Lincoln.

Wright, V.P. 1985. The precursor environment for vascular plant colonization. *Philosophical Transactions of the Royal Society of London* B309: 143–145.

Zahner, R. 1967. Refinement in empirical function for realistic soil-moisture régimes under forest cover. Pp. 261–274 in W.E. Sopper and H.E. Lull (eds.) *Forest Hydrology.* National Science Foundation Advanced Seminar Proceedings. University of Pennsylvania/Pergamon Press, Oxford.

Zinke, P.J. and R.L. Crocker. 1962. The influence of giant sequoia on soil properties. *Forest Science* 8: 2–11.

Index